T0205835

ESSENTIAL QUANTUM MECHANICS

Essential Quantum Mechanics

GARY E. BOWMAN
Department of Physics and Astronomy
Northern Arizona University

OXFORD
UNIVERSITY PRESS

OXFORD

UNIVERSITY PRESS

Great Clarendon Street, Oxford OX2 6DP
United Kingdom

Oxford University Press is a department of the University of Oxford.
It furthers the University's objective of excellence in research, scholarship,
and education by publishing worldwide. Oxford is a registered trade mark of
Oxford University Press in the UK and in certain other countries

First published 2008
Reprinted 2012

British Library Cataloguing in Publication Data
Data available

Library of Congress Cataloging in Publication Data
Data available

ISBN 978-0-19-922893-5

Printed and bound by CPI Group (UK) Ltd, Croydon, CR0 4YY

Contents

Preface

While still a relatively new graduate student, I once remarked to my advisor, Jim Cushing, that I still didn't understand quantum mechanics. To this he promptly replied: "You'll spend the rest of your life trying to understand quantum mechanics!" Despite countless books that the subject has spawned since it first assumed a coherent form in the 1920s, quantum mechanics remains notoriously, even legendarily, difficult. Some may believe students should be told that physics really isn't that hard, presumably so as not to intimidate them. I disagree: what can be more demoralizing than struggling mightily with a subject, only to be told that it's really not that difficult?

Let me say it outright, then: *quantum mechanics is hard*. In writing this book, I have not found any "magic bullet" by which I can render the subject easily digestible. I have, however, tried to write a book that is neither a popularization nor a "standard" text; a book that takes a modern approach, rather than one grounded in pedagogical precedent; a book that focuses on elucidating the structure and meaning of quantum mechanics, leaving comprehensive treatments to the standard texts.

Above all, I have tried to write with the student in mind. The primary target audience is undergraduates about to take, or taking, their first quantum course. But my hope is that the book will also serve biologists, philosophers, engineers, and other thoughtful people—people who are fascinated by quantum physics, but find the popularizations too simplistic, and the textbooks too advanced and comprehensive—by providing a foothold on "real" quantum mechanics, as used by working scientists.

Popularizations of quantum mechanics are intended not to expound the subject as used by working scientists, but rather to discuss "quantum weirdness," such as Bell's theorem and the measurement problem, in terms palatable to interested non-scientists. As such, the mathematical level of such books ranges from very low to essentially nonexistent.

In contrast, the comprehensive texts used in advanced courses often make daunting conceptual and mathematical demands on the reader. Preparation for such courses typically consists of a modern physics course, but these tend to be rather conceptual. Modern physics texts generally take a semi-historical approach, discussing topics such as the Bohr atom and the Compton effect. Formalism is minimized and description emphasized; the highly abstract mathematical and physical concepts of quantum mechanics remain largely untouched. There is thus a rather large gap to be bridged, and students in advanced courses may find that they must solve problems and learn new applications even while the framework of quantum mechanics remains unclear.

Neither popularization nor standard text, this book is intended to serve in a variety of settings: as a primary text in a short course, a supplementary text in a standard course, a vehicle for independent study, or a reference work. Knowledge of elementary calculus and basic complex analysis should provide sufficient mathematical background (a condensed discussion of these topics appears in Appendix A).

The book's modernity is reflected in its overall style and tenor, but also in some broad themes, such as the early and extensive use of Dirac notation, and the fact that neither wavefunctions nor the time-independent Schrödinger equation are granted privileged status. Another such theme is the adoption of the "statistical interpretation," a very useful and lucid way to understand how quantum mechanics works in actual practice. Because the statistical interpretation is really a broad framework rather than an interpretation *per se*, it is easily "imported" into other approaches as the student may find necessary.

Notable by their absence from the book are many standard topics, such as perturbation theory, scattering, and the Hydrogen atom. This is in keeping with a central motivating idea: that to properly understand the many and varied applications of quantum mechanics, one must first properly understand its overall structure. This implies a focus on fundamentals, such as superposition and time evolution, with the result that they may then be developed in a more detailed and explanatory style than in advanced texts.

Some authors seem to believe that if they provide a clear, elegant, terse explanation, one time, any remaining confusion is the student's responsibility. I disagree. Having taught (and learned) physics for many years at many levels, I find that there are myriad ways to *mis*understand the subject, so I have tried to make this book especially explanatory and useful for the student. Common variations in terminology and notation are clarified (e.g., the terms quantum state, state vector, and wavefunction). And I discuss not only what is right, but what is wrong. For example, although position-space and momentum-space are standard topics, students often fail to realize that there is but one quantum state, which may be cast into various representations. Such potential stumbling blocks are explicitly pointed out and explained.

The great majority of problems are, to my knowledge, new. Most are intended to help develop conceptual understanding. A vast array of additional problems may be found in other quantum texts. The time-honored physics dictum—that one doesn't understand the physics unless one can solve problems—bears repeating here. But so does its lesser-known cousin: *just* solving problems, without the capacity to lucidly discuss those problems and the attendant concepts and ideas, may also indicate insufficient understanding.

In part because this book is intended to transcend the traditional physics audience, a few words about studying the subject are in order. Much of our intellectual heritage–from art and music to social, political, and historical thought–concerns our human experience of the world. By its very nature, physics does not, and it is now clear that at the fundamental level the physical world doesn't conform to our preconceived ideas. The concepts of physics, particularly quantum mechanics, can be exceedingly abstract, their connections to our everyday experiences tenuous at best.

Because of this physical abstraction, and the requisite mathematical sophistication, understanding can be hard to achieve in quantum mechanics. Nevertheless, I believe that understanding (*not* memorization) *must* be the goal. To reach it, however, you may need to read more carefully, and think more carefully, than ever before. *This is an acquired skill!* For most humans it simply isn't natural to exert the degree of concentration that physics demands–you didn't think quantum mechanics would be easy, did you? The payoff for this hard work, to borrow Victor Weisskopf's phrase, is *the joy of insight.*

Essential Quantum Mechanics would not have become a reality absent the freedom and support granted me by Northern Arizona University. This includes a sabbatical spent, in part, developing the book at Loyola University Chicago. Professor Ralph Baierlein generously and critically read the manuscript and, as always, provided much wise and deeply appreciated counsel. Professor Peter Kosso offered useful comments and early encouragement. Sonke Adlung, of Oxford University Press, displayed abundant patience, kindness, and professionalism in helping me through the publishing process. Oxford's Chloe Plummer endured my repeated underestimates of the time required to correct the manuscript.

The influence of my late, great, Ph.D. advisor, Jim Cushing—whose life put the lie to the notion that scientists are not real intellectuals—permeates this book. My wife Katherine has been, and remains, a source of encouragement and forbearance through thick and thin. She also provided motivation, often by asking: When are you going to finish that #&!* book? Finally, I must thank my parents. Neither will see this book in print, yet both have indelibly impacted my life, and continue to do so, regardless of my age.

After more than a few years on the planet, it sometimes seems to me that there is but one great lesson to be learned. That is that the real worth of a life is in contributing to the welfare of others. It is my hope that, in some sense, and in some measure, I have done so with this book.

Flagstaff, Arizona *Gary E. Bowman*
May 2007

The true value of a human being is determined primarily by the measure and the sense in which he has attained liberation from the self.

Albert Einstein (1931)

1

Introduction: Three Worlds

The best things can't be told: the second best are misunderstood.
Heinrich Zimmer[1]

You may hear quantum mechanics described as "the physics of the very small," or "the physics of atoms, molecules, and subatomic particles," or "our most fundamental physical theory." But such broad, descriptive statements reveal nothing of the structure of quantum mechanics. The broad goal of this book is to reveal that structure, and the concepts upon which it is built, without becoming engulfed in calculations and applications. To give us something concrete to hold onto as we venture into the wilderness before us, and to give us a taste of what lies ahead, let's first take a little trip.

1.1 Worlds 1 and 2

Imagine a world; let's call it World 1. In World 1, everything is made up of very small, irreducible units called *particles*. (Large objects are composed of collections of these small units.) Because particles are the fundamental stuff of World 1, all physical events there are ultimately describable in terms of particle behavior—specifically, in terms of particle trajectories, the motion of particles in space as a function of time. Thus, to understand and predict events in World 1 we must understand and be able to predict the behavior of particles.

Our observations in World 1, then, are fundamentally observations of particle trajectories. Any association of physical properties with the particles, beyond their trajectories, is secondary—done to facilitate our understanding and predictive abilities. Nevertheless, it's convenient to postulate various measurable physical properties associated with the particles, and to give these properties names, such as mass and charge. (The definition and measurement of such properties may be a daunting task, but that is not our concern here.)

---

[1] Quoted in Campbell (1985), p. 21.

If these postulated physical properties are to be useful for understanding and predicting particle trajectories, we must construct a connection between the properties and the trajectories. This connection consists of two parts. First, we propose that the properties give rise to *forces*. In general, the connection between properties and the forces that they give rise to depends both upon the specific properties involved and upon the system's configuration—the positions and/or velocities of the particles. The forces are then connected to the particle trajectories by a set of dynamical laws. These dynamical laws, unlike the laws that give the forces themselves, are perfectly general: they connect *any force*, regardless of source, to the particle trajectories.

It is the job of the physicists on World 1 to define the physical properties of the particles, the way in which forces arise from these properties, and the dynamical laws which connect the forces to particle trajectories. And they must do so such that they obtain a consistent theoretical explanation for the particle trajectories—the fundamental observable entities of World 1.

The worldview of World 1 is, of course, that of Newtonian classical mechanics. In World 1, the complete description of a system consists of a description of the motions in time, that is, the trajectories, of all particles in the system. To obtain such a description, we determine the forces arising from the particles in the system by virtue of their various associated properties and the system configuration. Then, using very general dynamical laws—in classical mechanics, these are Newton's laws of motion—we connect forces with particle trajectories. Note that in World 1, as in Newtonian mechanics, no explanation is given of how forces are transmitted from one particle to another.

Now imagine another world: World 2. As in World 1, the tangible things of World 2 are made up of particles, and our goal is to determine and predict the trajectories of those particles. Now, however, the forces are transmitted from one particle to another by means of intangible *fields* which extend through space. In addition, dynamical properties, such as energy and momentum, are associated not only with the particles, but with the fields themselves.

The inhabitants of World 2 have found a simple mathematical algorithm such that if they know the fields, they can calculate the forces. Thus, from the particles and their associated properties they can find the fields, from the fields they can find the forces, and from the forces they can find the trajectories.

World 2 is the world of classical field physics. Here our starting point is the field (or equivalently, some potential from which the field is easily derived) created by some configuration of particles with their associated properties.

Note that in both World 1 and World 2 what we really observe are the particle trajectories. We never really "see" a field or a force, or even mass or charge—we only see their consequences, their effects on particle trajectories. Indeed, what we did was propose concepts such as mass, charge, and force as a means of providing a consistent theoretical base for our observations of particle trajectories, that is, *in response to* the observed trajectories.

1.2   World 3

We have one more stop: World 3. First, though, let's note that although the physicists on Worlds 1 and 2 may disagree on the rigorous definition of the word "particle," they would agree that we will rarely get into trouble if we think of a particle as a highly localized, even point-like, collection of physical properties.

In comparison to Worlds 1 and 2, World 3 is a very strange place indeed. The physicists on World 3 do not talk of particle trajectories. In fact, even though they constantly refer to particles, their conception of a particle seems vague at best, perhaps even muddled. Certain particle properties, such as mass and charge, are well defined in World 3, as in Worlds 1 and 2. But dynamical particle properties such as energy, momentum, and position, are quite poorly defined: their values seem to depend not only on the particle itself, but also on the sort of measurement we perform on it.

The key difference between Worlds 1 and 2 on the one hand and World 3 on the other, however, is the following. In World 2 the particle properties and the configuration of a system were used to determine fields, and from the fields, forces. (World 1 was similar except that the forces were determined directly.) The significance of the forces was their connection, by a dynamical law, to the time-dependent motion of the particles in the system.

In World 3, also, particle properties and the configuration of the system provide our starting point: together they give rise to potentials, as in World 2. Now, however, we do not make the transition to forces. Instead the particle properties and system configuration give rise to a probability function. This probability function only allows us to calculate the "chance" of finding a particle with a certain value for a dynamical property upon measurement. In Worlds 1 and 2 the dynamical law allowed us to calculate exactly the time-dependent motion of particles. In World 3 we also have a dynamical law; however, it only allows us to calculate the development in time of the probability function.

Although we still refer to particles in World 3, we may get into great difficulty if we imagine this to mean point-like units with well-defined trajectories. The "particles" of World 3 have poorly defined properties which only become well-defined upon measurement. Thus, we should not think of particle trajectories; that is, we should not think of a well-defined motion of the particles in space as a function of time. Only the probability function, which yields the statistics for obtaining certain values for the dynamical properties of the particles upon measurement, has a well-defined development in time.

The worldview of World 3 is essentially that of the most common interpretation of quantum mechanics. The contrast between Worlds 1 and 2 on the one hand and World 3 on the other illustrates, in large measure, the fundamental disparity between the classical and quantum worlds. This disparity arises at the deepest level. Particles and trajectories, the fundamental constituents of physical reality in the classical world, evidently are not well defined—if, indeed, they exist at all—in the quantum world.

1.3 Problems

1. An intelligent, thoughtful friend with essentially no mathematics or physics background reads this chapter.

   "In World 3," he says, "the probability function evidently plays a fundamental role. But I don't understand what a probability function *is*. When I asked a mathematician friend to explain, she just smiled and said "Oh, a probability function is just a probability distribution." But I don't know what that means, either."

   Please explain, in non-technical terms that your friend can understand, what a probability function, or distribution, is.

2. Consider World 3. In this world, what cannot be known precisely about a "particle"? Is there anything whose time evolution can be known precisely in World 3? If so, what is it?

3. This question and the next help emphasize the contrast between classical and quantum physics, by illustrating the overall structure of classical mechanics and classical field physics.

   The potential energy of a single, point particle of mass m is given by $U = \frac{1}{2}kx^2$, where k is a constant. We wish to find $x(t)$, the particle's motion as a function of time.

   (i) Clearly describe, step-by-step, in words, how you will find $x(t)$.

   (ii) Carry out the instructions you provided in part i.

   (iii) Now assume that at time $t = 0$ the particle is stationary at $x = \alpha$. Find the specific $x(t)$ for this case.

   (iv) Does any information about our point particle's motion remain unknown, or do we now know all of it?

   (v) What is the common name used to describe this sort of system?

4. Assume that the particle in the preceding question has charge q, and that its potential energy U arises from the presence of an electric field.

   (i) Find the particle's potential, V.

   (ii) What is the electric field in which the particle resides?

2

The Quantum Postulates

Quantum mechanics [is] that mysterious, confusing discipline, which none of us really understands but which we know how to use. It works perfectly, as far as we can tell, in describing physical reality, but it is a "counter-intuitive discipline," as the social scientists would say. Quantum mechanics is not a theory, but rather a framework within which we believe any correct theory must fit.[1]
Murray Gell-Mann

Quantum mechanics is often developed from postulates. I'll also follow this approach; it has the distinct advantage that the entire theoretical foundation of the theory can be written down in a few concise statements.

The postulate approach is subject to the criticism that it tends to emphasize the distinct natures of classical mechanics and quantum mechanics, and thus largely fails to illuminate the many similarities of the two theories. I will try to alleviate this deficiency somewhat in our discussion.

What might often confound the physics student are apparently dissimilar statements made regarding some topic in various references. Frequently it is only after much study and thought that one finally sees that the various statements are in fact equivalent. The quantum postulates are a good example of just such a case. Most quantum mechanics texts contain the postulates, but they may not explicitly call them postulates. In addition, although the *content* of the postulates is almost universal, there often seems to be almost as many ways to *state* them as there are authors of quantum mechanics texts.

Before we begin our examination of the postulates, it is worthwhile to contemplate what a postulate is. Consider Newton's laws of motion. An astonishing amount of physics has been developed using these three simple statements as their foundation. Linear dynamics, rotational dynamics, Lagrangian and Hamiltonian mechanics, gravitation, classical statistical mechanics, electrodynamics—all of these are grounded in Newton's laws. In a sense these are all, to some degree, *derived* from Newton's laws. Arguably, even quantum mechanics itself is ultimately grounded in Newtonian mechanics.[2]

---

[1] M. Gell-Mann, in Mulvey (1981).
[2] Dirac (1978), p. 2.

So, what are Newton's laws derivable from? The answer is: Newton's laws cannot be derived from anything. Any truly fundamental "law of nature" must be underivable. *That's what makes it a law of nature.* Such a law tells us something about nature "in the raw"; there is no way to derive it from either physical or mathematical considerations. Of course, we may find later that what we had previously thought was truly fundamental really isn't. But no matter: in that case, we simply find that underivability, that is, fundamentality, lies at a deeper level than we previously thought. (Some may claim that Newton's laws, for example, are not truly fundamental, but are derivable from quantum mechanics. If this view is correct then underivability lies at the quantum mechanical level, not at the classical mechanical level.)

A postulate is, essentially, a statement made without proof or analytical justification—an underived statement. In physics a postulate is, in effect, a proposal, and such a proposal must ultimately stand or fall not on *derivation*, but on *verification* in the physical world. It must be tested experimentally. Newton's laws of motion were such proposals. Quantum mechanics also makes such proposals: the quantum postulates. We will spend much of this book investigating these postulates and their consequences.

2.1 Postulate 1: The Quantum State

Information about a quantum-mechanical system is contained in the quantum state. The form and development in time of the state are determined by the time-dependent Schrödinger equation (TDSE):

$$i\hbar\frac{d\Psi}{dt} = \hat{H}\Psi \qquad (2.1)$$

where Ψ is the quantum state and $\hat{H}$ is the Hamiltonian operator.

We have some distance to go before we can fully grasp the meaning of Postulate 1. Even now, however, we may utilize it to begin to develop one of the most deceptively subtle, and disputed, concepts in quantum mechanics: the *quantum state*.[3] Elsewhere, you will often find quantum states referred to as *wavefunctions* or *state vectors*. For now, we will use these terms sparingly. The term quantum state is both more general and more descriptive, and it reminds us that the quantum state is the analog of the state of a classical system.

Let's first consider the state of a system at some particular time in Newtonian mechanics. In this case the state of the system is given by the positions and velocities of all its constituent parts. From this information, and from the forces which determine the particles' subsequent behavior, we

---

[3] Ballentine (1998), Section 2.1.

may (in principle) use Newton's laws to determine the state of the system at any later time. The concept of state for a classical system is thus almost intuitively clear.

Now what does Postulate 1 tell us about the quantum state? As stated, $\hat{H}$ is the Hamiltonian *operator*. In quantum mechanics, operators are usually denoted by a "hat". We will develop operators in more depth later, but for now it is sufficient to state that $\hat{H}$ is not just a number, or even a function. Rather it involves mathematical operations, such as differentiation.

From Eq. (2.1), the TDSE clearly is not just an *algebraic* equation, but a *differential* equation.[4] The solution to a differential equation is a *function*—in general, a *complex* function. Consider, for example, the (very simple) differential equation:

$$\frac{d^2 f(x)}{dx^2} + k^2 f(x) = 0, \tag{2.2}$$

where k is a real constant and $f(x)$ is the function which is a solution of the differential equation. (That is, $f(x)$ is what we must find to solve the differential equation.) It is easily verified that the complex function,

$$f(x) = e^{ikx} = \cos(kx) + i\sin(kx), \tag{2.3}$$

is a solution of Eq. (2.2). In Postulate 1, the solution to the TDSE is Ψ. Evidently Ψ is a function of time. In general, Ψ is complex, and will remain so as time evolves (progresses).

Moreover, quantum states are not simply specifications of the corresponding classical quantities. Consider a single *classical* particle. The state of a classical particle at some time is given by the three position coordinates and the three velocity components of the particle. It turns out, however, that the *quantum* state of a single particle at a particular time is, in general, a *function* in space. This means that, in contrast to the classical state, we can't simply specify unique coordinates and velocities of a quantum particle. This is a mathematical manifestation of the difference between Worlds 1 and 2, on the one hand, and World 3 on the other (as described in Chapter 1).

Finally, note that Postulate 1 states that "information about a quantum-mechanical system is contained in the quantum state." But could information also reside somewhere *other* than in the state? Elsewhere you may read statements such as "*all* information resides in the quantum state," or "the *most complete* information resides in the quantum state." In this view, the quantum state provides a *complete* description of a quantum-mechanical system. Postulate 1, however, leaves open the possibility that

---

[4] Often $\hat{H}$ involves partial differentiation; then the TDSE is a *partial* differential equation.

the quantum state may not be the whole story—that there could be some deeper layer of physics beyond quantum mechanics. These two views thus entail profoundly different philosophical perspectives. Indeed, the debates in foundations of quantum mechanics, still ongoing after 80 years, largely spring from the difference in these perspectives. We will return to the issue of the completeness of quantum mechanics in Chapter 3.

Although Postulate 1 is absolutely essential for the construction of quantum mechanics, it is virtually devoid of *physical* content. It tells us something about the quantum state and how to find it, but nothing about how to connect the quantum state to physically observable predictions. We begin building that connection in the next section.

2.2 Postulate 2: Observables, Operators, and Eigenstates

Properties of a quantum-mechanical system that can (in principle) be observed, or measured, are called *observables*, and are represented by Hermitian operators. Suppose the quantum state Ψ_j satisfies:

$$\hat{A}\Psi_j = a_j\Psi_j, \tag{2.4}$$

where $\hat{A}$ is an operator representing an observable, and a_j is a constant. Then Ψ_j and a_j are the jth *eigenstate* and *eigenvalue*, respectively, of $\hat{A}$. If the system is in the state Ψ_j, then the result of a measurement of the observable represented by the operator $\hat{A}$ *must* be the eigenvalue a_j.

Postulate 2 tells us that measurable properties are called observables in quantum mechanics, and that, in the mathematical formulation of the theory, observables are represented by Hermitian operators. But what do these statements mean? To start to answer that, let's first briefly revisit classical mechanics. If we consider a single particle system, we see that once we know the trajectory of the particle, $\vec{x}(t)$, that is, its position as a function of time, we can calculate the other physical properties that we're usually interested in.[5] For example, the momentum, $\vec{p} = m\vec{v}$, is found by multiplying the particle's mass by its velocity, where the velocity is found by differentiating the trajectory $\vec{x}(t)$. A similar procedure can be used to find, for example, the angular momentum about some point. In other words, once we have calculated the particle's trajectory we may simply apply the *definitions* of various physical properties to calculate the values of such properties.

In quantum mechanics, however, we do not have well-defined particle properties to work with. The fundamental entity available to us is not the

---

[5] Here $\vec{x}$ denotes the position vector (x, y, z) or, in another common notation, (x_1, x_2, x_3).

particle trajectory, but the quantum state—a function that is, in general, "spread out," or distributed, in space, and from which we obtain probabilities. We certainly cannot simply apply, say, the definition of linear momentum, $\vec{p} = m\vec{v}$, to the quantum state and hope to get out the linear momentum of the particle. If we tried to do so, we would immediately be faced with the question of what to calculate the velocity *of* . . . the quantum state?

Postulate 2 prescribes the method to resolve these questions. Observables will be represented (mathematically) in the theory by Hermitian operators. We will develop later what a Hermitian operator is, along with other important properties of the operators commonly used in quantum mechanics. The important point now is that an operator is, in general, not just a scalar, a vector, or a function. It is a mathematical operation, such as differentiation with respect to spatial coordinates.

Postulate 2 then tells us that if a measurement of the observable corresponding to $\hat{A}$—call it A—is performed on Ψ_j, the jth eigenstate of $\hat{A}$, the result will be the eigenvalue a_j. Let's unpack this statement a bit. The equation of Postulate 2,

$$\hat{A}\Psi_j = a_j\Psi_j, \tag{2.5}$$

is an *eigenvalue equation*. On the left side of Eq. (2.5) the operator $\hat{A}$ acts on the function Ψ_j. (Remember that $\hat{A}$ is a mathematical operation, not just a number or a function.[6]) The right side then tells us that this equals the eigenvalue, a_j, multiplied by Ψ_j. An eigenvalue is *not* an operator; it is simply a *number*.

Now let me state the physical interpretation of Eq. (2.5). Consider a system in the quantum state Ψ_j. Then if I measure A, the observable corresponding to the operator $\hat{A}$, the measured value will be the eigenvalue (number) a_j. Let's say I wish to measure linear momentum, which corresponds to the operator $\hat{p}$. Suppose the system is in the state ϕ_k, the k'th eigenstate of $\hat{p}$. Then the relevant eigenvalue equation is

$$\hat{p}\phi_k = p_k\phi_k. \tag{2.6}$$

Our measurement result will be the number p_k, an eigenvalue of $\hat{p}$. I don't want to delve deeper into eigenvalues and eigenstates at this point. I simply want you to begin developing some "feel" for how eigenvalues, eigenstates, and eigenvalue equations enter quantum mechanics.

Please realize that *the eigenstates of an operator* and *the state of a physical system* are distinct concepts. In discussing both Eqs. (2.5) and (2.6), as in Postulate 2, I stipulated that the state of the system was *also* an eigenstate of the relevant operator. However, for any quantum-mechanical observable there are infinitely many quantum states that *are* legitimate

---

[6] If, for example $\hat{A}$ is a differential operator, then Eq. (2.5) is a differential equation.

states of the system, but are *not* eigenstates of the relevant operator. For such state/operator combinations, eigenvalue equations such as Eqs. (2.5) and (2.6) do *not* apply.

For example, if ψ is *not* an eigenstate of $\hat{A}$, no number β exists such that $\hat{A}\psi = \beta\psi$ is satisfied. *An eigenvalue equation for some operator, say $\hat{A}$, holds only if the function (state) on which $\hat{A}$ acts is an eigenstate of $\hat{A}$.* A very important result, then, is that *an operator, a mathematical operation, may be replaced by the corresponding eigenvalue, a number, but only if the state on which the operator acts is an eigenstate of the operator.*

What is the physical import of eigenstates? As we shall see, quantum states yield probabilities for measurements of observables. But Postulate 2 tells us that the probability distribution for an observable is trivial *if* the quantum state is an eigenstate of the corresponding operator. For the state Ψ_j, for example, the probability distribution for measurement of A is trivial: the probability of obtaining a_j upon measurement is 1, and the probability of obtaining any other value is 0. We will, *with certainty*, obtain a_j. For states that are *not* eigenstates of $\hat{A}$, however, the probability distribution will be non-trivial, and that is what our last postulate deals with.[7]

2.3 Postulate 3: Quantum Superpositions

If a measurement of the observable corresponding to the operator $\hat{A}$ is made on the normalized quantum state ψ, given by,

$$\psi = \sum_n c_n \Psi_n, \qquad (2.7)$$

where the Ψ_n's are eigenstates of $\hat{A}$ and the c_n's are expansion coefficients, then a_j, the eigenvalue of Ψ_j, will be obtained with probability $|c_j|^2$. The system will be left in the state Ψ_j immediately after the measurement.

Here $|c_j|^2$ is the *complex square* of c_j, that is, $|c_j|^2 = c_j^* c_j$ (see Appendix A).

In a sense, Postulate 3 simply extends Postulate 2 to cases where the state of the system is not an eigenstate of the operator corresponding to the observable being measured. Nevertheless, the *physical* implications of Postulate 3 have probably given rise to more controversy than any other aspect of the foundations of quantum mechanics. Our discussion of Postulate 3 is not intended to be rigorous, but rather to be sufficiently descriptive that you may begin to form a mental picture of the structure of quantum mechanics.

Before developing Postulate 3 proper, we define a *normalized* quantum state. Normalization simply accounts for the physical requirement that the probabilities for all possible results of some type of measurement must add

---

[7] The concept of probability is discussed in detail in Chapter 3.

up to one. In the context of Postulate 3, normalization amounts to imposing the condition:

$$\sum_n |c_n|^2 = 1. \tag{2.8}$$

If a state is not normalized, we impose normalization by multiplying all coefficients by a constant such that Eq. (2.8) is obeyed.

2.3.1 Discrete Eigenvalues

Suppose a system is in the quantum state Ω, and I wish to calculate the results I would obtain if I measured either of two observables, A or B, corresponding to the operators $\hat{A}$ and $\hat{B}$, respectively. Specify that eigenstates of $\hat{A}$ are *not* eigenstates of $\hat{B}$, and vice versa. Moreover, the state of the system, Ω, need not be an eigenstate of either $\hat{A}$ or $\hat{B}$.

I'll call the normalized eigenstates of $\hat{A}$ the Ψs, and those of $\hat{B}$ the ϕs. It turns out that the set of all normalized eigenstates of a Hermitian operator, such as the set of all Ψs, or all ϕs, forms a *complete set* of states. I will develop the critical concept of a complete set of states in considerable detail in Chapter 4—the important point for now is that any state may be "expanded" in terms of such a set. That is, any state may be written as a linear combination—a weighted sum—of such states. Thus, Ω may be expanded as a linear combination of either the Ψs or the ϕs:

$$\Omega = \sum_{m=1}^{N} c_m \Psi_m,$$

$$\Omega = \sum_{k=1}^{N} \alpha_k \phi_k, \tag{2.9}$$

where the summations run over all N states that comprise the complete set, and the c_ms and α_ks are called *expansion coefficients*. In quantum mechanics, a state which is expanded as a linear combination of (more than one of) the eigenstates of some operator is usually referred to as a *superposition state*, or just a *superposition*.

Let's look at the case for the observable B. The eigenvalue equation is

$$\hat{B}\phi_j = b_j\phi_j, \tag{2.10}$$

where the b_js are the eigenvalues. But how do we deal with a state that is not an eigenstate of $\hat{B}$? For example, what do we do with $\hat{B}\Omega$? We need only expand Ω as a superposition of the eigenstates of $\hat{B}$ to see that

$$\hat{B}\Omega = \hat{B}\sum_{k=1}^{N} \alpha_k \phi_k = \hat{B}(\alpha_1\phi_1 + \alpha_2\phi_2 + \cdots + \alpha_N\phi_N)$$

$$= b_1\alpha_1\phi_1 + b_2\alpha_2\phi_2 + \cdots + \alpha_N\phi_N. \tag{2.11}$$

Well, we can write this out formally, but what does it *mean*? Postulate 2 tells us, through eigenvalue equations, about measurements, but only on *eigenstates* of the relevant operator, not on *superpositions* of eigenstates of that operator. So, while Eq. (2.11) is mathematically correct, it doesn't predict the results of measurements of the observable corresponding to $\hat{B}$ on the state Ω.

Postulate 3 tells us how to handle such cases. The result of any single measurement of the observable corresponding to $\hat{B}$ on the system state Ω will be just *one* of the eigenvalues of $\hat{B}$, not a "superposition" of eigenvalues. The probability of obtaining, say, the jth eigenvalue of $\hat{B}$, denoted $Prob(b_j)$, is given by the complex square of the expansion coefficient of the jth eigenstate of $\hat{B}$. That is, $Prob(b_j) = \alpha_j^* \alpha_j = |\alpha_j|^2$. Postulate 3 thus provides the means to construct a probability distribution for measurement of *any* observable with the system in *any* quantum state.[8] Finally, Postulate 3 tells us that, after a measurement yields a particular eigenvalue, the system is left in the corresponding eigenstate.

Simple as it is, Postulate 3 is one of the more remarkable statements in physics. It says that even if the system state cannot be characterized by only one eigenvalue, we will nevertheless obtain only one eigenvalue upon measurement—this is the root of the infamous *measurement problem* of quantum mechanics. And although Postulate 3 provides a rule for determining the probability of obtaining each eigenvalue, it provides no explanation as to the physical source of these probabilities. Finally, we could hope to argue that these probabilities simply reflect changes in our state of knowledge, not in the physical system. But the postulate forecloses such attempts, by stating explicitly that the physical state of the system is indeed altered in accord with the measurement result.

2.3.2 Continuous Eigenvalues

To simplify things, I've limited our discussion to operators with discrete eigenvalues. The continuous eigenvalue case (which we will discuss in detail in Chapter 12) tends to be less transparent, and involves mathematical subtleties not essential to understanding the overall quantum framework. In both quantum and classical physics, position may take on a continuum of values. This is reflected in that particular type of quantum state often called a wavefunction. Because introductory students often have some familiarity with wavefunctions, but not with the more general concept of a quantum state, I now establish the connection between wavefunctions and our postulates.

Although it's not obvious, the following statement essentially combines Postulates 2 and 3 and restates them in terms of wavefunctions.

---

[8] Postulate 3 must be modified for use with *mixed states*, also called mixtures. The mixed state concept is, however, rather advanced; I will not discuss it.

Let $\Psi(x)$ be the normalized wavefunction (quantum state) of a single-particle system. Then the interpretation of $\Psi(x)$ is that the probability of finding the particle, upon measurement, to be in the infinitesimal interval between x and $x + dx$ is $|\Psi(x)|^2 dx$. Equivalently, the probability of finding the particle in the finite interval (a, b) upon measurement is

$$\int_a^b |\Psi(x)|^2 dx. \tag{2.12}$$

The particle will be left in the eigenstate corresponding to the measured value of x immediately after the measurement.

In Postulate 3, the $|c_j|^2$s were *probabilities*. But $|\Psi(x)|^2$ is evidently a probability *density* (of dimensions probability/length), which must be *integrated* to obtain a finite probability. This is similar to, say, calculating the total mass in some region of space. If the mass in the region comprises discrete objects, we simply sum their individual masses. But if the mass is continuously distributed, we must *integrate* the mass *density* (of dimensions mass/length^3).

Normalization now accounts for the physical requirement that, upon measurement, the particle must be found *somewhere*. This implies that if the probability density is integrated over the entire x axis, the result must be 1:

$$\int_{-\infty}^{\infty} |\Psi(x)|^2 dx = 1. \tag{2.13}$$

Establishing a connection with the postulates hinges on three key points (justified in Chapter 12). First, each unique position, such as x_0, corresponds to a unique eigenstate of the position operator $\hat{x}$; the corresponding eigenvalue is just x_0 itself. Second, the expansion coefficient for the eigenstate corresponding to x_0 is $\Psi(x_0)$, that is, the wavefunction evaluated at x_0. Finally, for a wavefunction, such as $\Psi(x)$, the operator $\hat{x}$ corresponding to the position observable is simply x itself (that is, $\hat{x} = x$ in this case).[9]

Now suppose a system is in the eigenstate corresponding to the eigenvalue x_0. Then the only non-zero coefficient, also, corresponds to x_0; that is, $\Psi(x) = 0$ for $x \neq x_0$. Thus, if a position measurement is made, the result will with certainty be x_0. This is in accord with Postulate 2, which states that if a system is in the jth eigenstate of $\hat{A}$, then measurement of A must yield the eigenvalue a_j.

What if the system is *not* in a position eigenstate? The expansion coefficient for each eigenvalue (each x) is $\Psi(x)$ itself, while the probability density is $|\Psi(x)|^2$. This is in accord with Postulate 3, which states that if

---

[9] This holds only because wavefunctions are generally written in the position representation—I discuss representations in Chapter 4.

a system is in a superposition of the eigenstates of $\hat{A}$, the probability of obtaining the jth eigenvalue, a_j, is the square of the jth coefficient, $|c_j|^2$.

In case you are mystified by the complex square, I will write out a simple, specific example for you. Assume the quantum state is given by

$$\Psi(x) = Ae^{ikx}, \tag{2.14}$$

where A is a (possibly complex) constant and $i = \sqrt{-1}$. Then

$$|\Psi(x)|^2 = \Psi(x)^*\Psi(x) = A^*e^{-ikx}Ae^{ikx} = |A|^2. \tag{2.15}$$

Upon measurement, only one x value will result, but for this simple state all xs are equally likely to be found.

Equation (2.12) states that the probability that measurement will find the particle in the interval (a, b) is obtained by adding up (integrating) the probabilities that the particle will be found at each point within that range. The same idea holds in Postulate 3: the probability that, say, either a_1 or a_3 will be found upon measurement of A is just the sum $|c_1|^2 + |c_3|^2$.

An important complication that arises in connection with continuous eigenvalues is the nature of the eigenstates, which are both mathematically abstract and physically unrealizable. Because this topic requires some mathematical sophistication to handle properly, I will simply sketch it here; a more detailed discussion appears in Chapter 12.

I have written eigenvalue equations of the form,

$$\hat{p}\,\phi_k = p_k\phi_k. \tag{2.16}$$

This is an example of an eigenvalue equation for the case of discrete eigenstates and eigenvalues. But what about cases in which the eigenvalues and eigenstates form a continuum? For example, upon measurement, a particle could be found to be anywhere along the x axis. Thus, there must exist a continuous set of position eigenvalues, the xs, and the position eigenvalue equation is,

$$\hat{x}\Psi = x\Psi. \tag{2.17}$$

The lack of subscripts implies the continuous nature of the eigenvalues. The probability is given by the continuous probability density: $|\Psi(x)|^2$.

Accompanying the continuous eigenvalues is a continuous set of eigenstates.[10] Such states are described by *Dirac delta functions*, denoted $\delta(x)$. The basic properties of $\delta(x)$ are

$$\begin{aligned} \delta(x-a) &= 0 \quad \text{if } x \neq a \\ \int_{x_1}^{x_2} \delta(x-a)dx &= 1 \quad \text{if } x_1 < a < x_2. \end{aligned} \tag{2.18}$$

---

[10] Note that we are not referring here to the continuity of the eigenstates themselves, but rather to the continuity of the *set* of eigenstates. That is, rather than a discrete set of eigenstates, such as $\Psi_1, \Psi_2, \ldots, \Psi_n$, the set of states is continuous, so that it doesn't make sense to give the states individual labels such as $1, 2, \ldots, n$.

The above is to be interpreted as meaning that $\delta(x-a)$ vanishes everywhere except at the point where its argument vanishes, and that at that point the *integrated* value of $\delta(x-a)$ is 1.[11] Since the eigenstates of position are Dirac delta functions, the position eigenvalue equation for, say, x_0, may be written as

$$\hat{x}\delta(x-x_0) = x\delta(x-x_0) \quad \longrightarrow \quad \hat{x}\delta(x-x_0) = x_0\delta(x-x_0). \qquad (2.19)$$

The second equation is valid because $\delta(x-x_0) = 0$ except at x_0.

If the Dirac delta function seems tricky, that's because it is. In fact, the Dirac delta is not a proper mathematical function, and only many years after it was first used by physicists did mathematicians deem such use legitimate.

2.4 Closing Comments

It should now be clear that in quantum mechanics only probabilistic predictions are generally possible (the exception being if the state is an eigenstate of the relevant operator). But consider, for example, a wavefunction such as $\Psi(x)$. The probabilities are obtained not from $\Psi(x)$ itself, but from $|\Psi(x)|^2$, so that it is through $|\Psi(x)|^2$ that quantum mechanics makes contact with the physical world. So why not just work with $|\Psi(x)|^2$, and call *that* the quantum state? What's the big deal?

The big deal is that if the fundamental thing is taken to be $|\Psi(x)|^2$, rather than $\Psi(x)$, *you absolutely won't have quantum mechanics any longer!* One way to see this is that the time-dependent Schrödinger equation—a fundamental law of nature—is a differential equation for Ψ, not for $|\Psi|^2$. Moreover, a *complex* function such as Ψ must, in general, embody more information than its real progeny, $|\Psi|^2$. Remember, *we square Ψ only upon measurement*. Absent a measurement, we must work with Ψ itself, not $|\Psi(x)|^2$.

Before closing, a few words about terminology are in order. In classical physics, the displacement of a wave (or, often, its *maximum* displacement) is called its *amplitude*. In analogy, quantum-mechanical wavefunctions—and, by extension, all quantum states—are often called *probability amplitudes*.

The square of a classical wave's amplitude is of great physical interest, because it determines the rate at which the wave transmits energy. Similarly, probabilities are of great physical interest in quantum mechanics, and they are obtained by (complex) squaring probability amplitudes, that is, quantum states.

---

[11] A concise summary of important properties of the Dirac delta may be found in *Jackson* (1999), pp. 26–27.

The term *superposition*, also, is rooted in classical wave theory, wherein it refers to a sum of individual wave amplitudes. By analogy, a superposition state (whether for continuous or discrete eigenvalues) is a quantum state that's expanded as a linear combination—a weighted sum—of the eigenstates of some operator.

A final note about terminology: I've been careful about referring to "the operator corresponding to some observable," and vice versa. Such terminology is correct, but cumbersome, so authors sometimes don't explicitly distinguish between observables and their corresponding operators. For example, you may see a phrase such as "measuring an operator," even though what's really measured is the observable corresponding to an operator. Please keep in mind that the terms operator and observable are not quite synonomous.

Well, that's it. Although I've certainly left some gaping holes in the fabric, and I'll keep trying to fine tune our understanding of the postulates as we proceed, I've now laid out the basic structure that forms the foundation for all of quantum mechanics.

Let me end by briefly summarizing where we've been. The fundamental entity in quantum mechanics is the quantum state, a time-dependent function that is obtained by solving the time-dependent Schrödinger equation (with the appropriate Hamiltonian operator). Observable quantities are represented by Hermitian operators, and the result of a measurement of some observable must be one of the eigenvalues of the corresponding Hermitian operator. If the system is in an eigenstate of the operator when measured, the result will, with certainty, be the eigenvalue corresponding to that eigenstate. If the system is in a superposition with respect to the measurement operator, the probability of obtaining any particular eigenvalue is the complex square of the expansion coefficient for the corresponding eigenstate.

Innocuous though it may seem, the preceding paragraph describes one of the great creations of the human intellect. Attending that creation are some of the most profound and puzzling questions ever to have confronted that intellect.

2.5 Problems

1. Suppose an observable quantity B corresponds to the operator $\hat{B} = \frac{-\hbar^2}{2m}\frac{d^2}{dx^2}$. For a particular system, the eigenstates of this operator are

$$\Psi_n(x) = A\sin\frac{n\pi x}{L}, \quad \text{where: } n = 1, 2, 3, \dots$$

Here A is a normalization constant.
(i) Determine the eigenvalues of $\hat{B}$ for this case.
(ii) Determine the probabilities, and the corresponding measured values, for B if the system is in the state $\Phi(x) = 0.8i\,\Psi_2(x) + 0.6\,\Psi_4(x)$.

2. The ket $|x'\rangle$ represents the position eigenfunction corresponding to the position x'. Write down the position-space representation of $|x'\rangle$.
3. A wavefunction $\psi(x)$ constitutes a set of (choose one):
   (i) expansion coefficients (probability amplitudes);
   (ii) probabilities;
   (iii) position operators.
4. Consider some momentum-space wavefunction $\phi(p,t)$ (analogous to the position-space wavefunction $\Psi(x,t)$ in Section 2.3.2). What is the probability that a measurement of momentum at time t' would result in a value between p_0 and $p_0 + \Delta p$? (*Hint*: just write down an integral.)
5. You have a lab in which you can create a "particle in a box". By this we mean a particle that is subject to the potential:

$$U = \infty, \quad x < 0 \text{ and } x > L,$$
$$U = 0 \text{ elsewhere.}$$

Note that any quantum state vanishes identically for $U = \infty$. The Hamiltonian operator $\hat{H}$, which appears in the Schrödinger equation, is also the operator corresponding to energy. The *unnormalized* energy eigenstates for this system are given by

$$\psi_n(x,t) = \sin\left(\frac{n\pi x}{L}\right)\exp(-iE_nt/\hbar) \quad n = 1, 2, 3, \ldots, \infty.$$

Here E_n is the nth energy eigenvalue (you will not need the explicit form of the E_ns for this problem).
(i) Determine the proper normalization factor (a constant) for these states.
(ii) Sketch the potential, U. Then sketch the modulus of ψ_1 (see Section 10.1.2) and $Prob(x)$ for ψ_1.
(iii) Repeat Part (ii) for ψ_2.
(iv) How does $Prob(x)$ depend on time for ψ_1 and ψ_2?
(v) You have a "black box" in your lab, a state preparation device, that does a pretty good job of preparing the initial state of the system in the form of a rectangularly-shaped packet of width $L/2$, centered at $x = L/2$.
(vi) Calculate the first five expansion coefficients (i.e. for $n = 1, 2, 3, 4, 5$) for this state.
(vii) Provide a clear, qualitative argument as to why the expansion coefficients for $n = 2$ and $n = 4$ behave as they do.
(viii) Denote the superposition of the $n = 1$, $n = 2$, and $n = 3$ states $\phi_3(x,t)$, and of the $n = 1$ through $n = 5$ states as $\phi_5(x,t)$.
(a) Show explicitly that ϕ_3 and ϕ_5 are not properly normalized.
(b) Properly normalize ϕ_3 and ϕ_5.

(c) Clearly argue why ϕ_3 and ϕ_5 were not automatically normalized. (This will probably require some careful thinking!)

(ix) Use a plotting program to plot $Prob(x)$ for $\phi_3(x, t = 0)$; then do the same for $\phi_5(x, t = 0)$.

(x) What do you expect to happen as the number of terms in the superposition increases (i.e. as we include more ψ_ns)? Can you see evidence for this in comparing ψ_1, ϕ_3, and ϕ_5?

(xi) We have discussed discrete eigenvalues, and in Chapter 2 we also discussed continuous eigenvalues a bit. Suppose a physics-student friend asks you, "Are the eigenvalues of ϕ_3 and ϕ_5 discrete or continuous?" You begin your answer with the statement "It depends on which eigenvalues you're referring to." Finish your answer by clearly explaining this statement to your friend.

6. To answer the following questions, set up and solve the eigenvalue equation for the operator corresponding to momentum along the x axis, that is: $-i\hbar\frac{d}{dx}$.

(i) What are the eigenstates (i.e. functions of x) and eigenvalues of this operator. Is the set of eigenfunctions and eigenvalues discrete or continuous?

(ii) Write down a superposition of two of the momentum eigenstates. Be sure that your superposition is normalized, that is, that the total probability equals 1. This condition arises from the fact that if you make a measurement—in this case, of momentum—you must get *something* as a result.

(iii) What is the probability distribution for momentum for your superposition?

3
What Is a Quantum State?

The assumption... is that if a student learns to solve problems, the proper under-standing will come naturally as he does so. I have met a number of intelligent physicists for whom this simply is not so—having in the course of time lost hold of the computational devices, they have lost all, and no instinct remains to guide them.[1]
David Park

What does the formalism of quantum mechanics mean? How is that formalism related to the physical world? What is the meaning of the quantum state? These are deep questions, and although they are addressed in both popular discussions and in the vast literature on the foundations of quantum mechanics, in many ways they remain unanswered.

Much of the purpose of this chapter—and indeed of this book—is to provide a framework for quantum mechanics that leads to a coherent, clear, and durable understanding. Somewhat surprisingly, this *can* be done; arguably, it is *not* the approach usually taken. In this chapter I first introduce some simple but essential concepts of probability and statistics. These are then used to develop a particularly lucid approach to quantum mechanics, the *statistical interpretation*, which we will adopt for the rest of our quantum journey. I close with a brief detour of the historical and philosophical milieu that was spawned by the sort of fundamental questions posed above.

3.1 Probabilities, Averages, and Uncertainties

3.1.1 Probabilities

Probability is the fundamental stuff of quantum mechanics. Why? Because quantum-mechanical predictions *must* take the form of probabilities—we generally cannot predict the outcome of a quantum-mechanical experiment with certainty. The end-goal of many quantum-mechanical calculations is, therefore, to obtain *probabilistic* predictions for the results of mea-surements. To understand quantum mechanics, then, you'll need a little knowledge of probability and statistics—but only a little.

---

[1] Park (1964), p. vi.

Since the quantum world often behaves in strange and unexpected ways, so do quantum probabilities. But although the *behavior* of quantum probabilities may be puzzling, their *statistical meaning* is precisely the same as that of "garden-variety" probabilities, such as the probability of obtaining "heads" in a coin toss, or of purchasing a winning lottery ticket. Only a few basic concepts of probability and statistics will be needed to develop quantum mechanics. If you understand these concepts in simple, unmysterious applications, you understand them in quantum mechanics. The quantum world is subtle enough as it is—don't make it more so by investing these concepts with undue mystery just because they appear in the context of quantum mechanics.

Probabilities are the raw material out of which other statistical concepts, such as averages and uncertainties, are built. But what *is* a probability? We need not delve into foundational issues here;[2] but we do need a practical, working definition.

Consider the following simple situation. A die has six faces. On each face are from one to six dots, each number of dots appearing only once. If I roll the die, what is the chance of obtaining a particular number of dots, say three? Because there appears to be no reason for the die to land preferentially on any face, I take each possibility as equally likely. I thus predict the chance of obtaining three dots on any roll to be $\frac{1}{6}$.

What I've called the chance is really the probability; clearly, any probability must lie in the interval $[0, 1]$. Suppose G is some measurable quantity, G_0 is a particular value of G, and $Prob(G_0)$ is the probability of obtaining G_0. If $Prob(G_0) = 0$, then G_0 never occurs; if $Prob(G_0) = 1$, G_0 occurs with certainty (always).

How do we know that the predicted probability in our example, $\frac{1}{6}$, is accurate? We test it experimentally, by repeatedly rolling the die. Say we roll it N times. If our prediction holds, we will find that as N becomes large, we will obtain three dots on approximately $\frac{1}{6}$ of the rolls, and this approximation will generally become better as N increases. In fact, as $N \to \infty$ we expect that the number of times we obtain three dots will approach exactly $N/6$, in accord with the predicted probability.

Clearly, the probability of obtaining any one of the six possible dot configurations is $\frac{1}{6}$. If we add up all of these probabilities—that is, if we take account of all six possible dot configurations, and thus all six possible results of a roll—we obtain a total probability of 1. This makes perfect sense: The probability that we will obtain *some* configuration on any roll—that *something* will happen—must be 1. This simple idea, which is rooted in probability (not in quantum mechanics *per se*), is the rationale for normalizing the quantum state.

In this example we used our knowledge of the situation to predict the probability of obtaining three dots. We then tested our prediction

---

[2] Such issues are briefly discussed in Ballentine (1998), pp. 31–33.

experimentally. Instead we could have simply run our experiment first, and then used the experimental results to surmise that the probability of obtaining three dots is $\frac{1}{6}$. We can connect these two approaches formally.

Suppose J is one possible outcome for each run of some experiment. I run the experiment N times, with N a very large number. If J is obtained N_J times, then I might write $Prob(J) = N_J/N$. Really, though, the experimental determination of a probability would require that $N \to \infty$ (which is, of course, unobtainable in practice). Then I would have

$$\lim_{N \to \infty} N_J/N = Prob(J), \tag{3.1}$$

which connects my experimental results with the idealized concept of a probability.

The collection of all probabilities associated with the various possible results of a measurement is the *probability distribution*. In the example of the rolled die, there are six distinct possibilities: the distribution is discrete. Suppose, however, that I also made repeated measurements of the *time* required for the die to come to rest after I released it. We would expect that these times could take on a continuous range of values. Suppose the *predicted* probability distribution for time, denoted $P(t)$ (with dimensions *probability/time*), takes the form of a Gaussian

$$P(t) \sim \exp^{-(t-t_0)^2}. \tag{3.2}$$

(Note that t_0 is the most probable time.) This is simply an assumption,[3] but a useful one: Gaussians possess nice mathematical properties, and they arise frequently in mathematics and physics.

Figure (3.1) illustrates both the discrete distribution for the number of dots obtained in a die roll, and the continuous distribution $P(t)$. For simplicity I will frame much of our discussion of quantum mechanics in terms of discrete distributions, but in Chapter 12 I will discuss continuous distributions in detail.

Finally, you should be aware of something that can cause much unnecessary consternation. Presumably, if we had sufficient information about any given roll of the die, and sufficient calculational ability, we could predict with certainty the dot configuration resulting from a roll. Thus, our use of probability really just reflects our lack of knowledge, our ignorance.

You may read elsewhere that things are much different in quantum mechanics, that quantum probabilities are not just reflections of our ignorance, but are instead deep, mysterious, and fundamental, revealing the operation of chance at nature's deepest levels—but is this true? This question is simply unanswerable at present (despite what anyone may claim). However, *you need not worry about such questions to learn quantum*

---

[3] This can't be exact, because it allows for $t < 0$, which is physical nonsense.

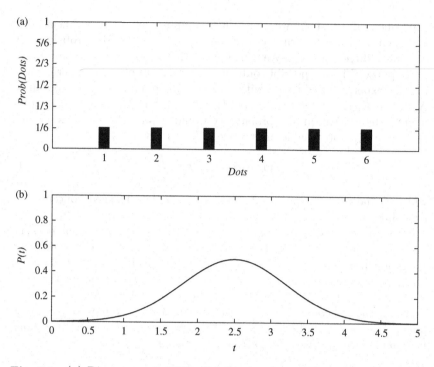

Fig. 3.1 (*a*) Discrete probability distribution for number of dots obtained in a die roll; (*b*) continuous probability distribution for time required to stop rolling. Here the most probable time, t_0, is 2.5.

mechanics. Regardless of any metaphysical meaning attached to quantum probabilites, the fact is that, in an operational sense, they are *just* probabilities—not unlike the probability of obtaining three dots when a die is rolled.

3.1.2   Averages

The notion of *average*, in the sense of "typical", is familiar from everyday life. But we'll need a quantitative definition. Fortunately, this definition is fairly transparent, and (unlike some mathematical or scientific terms) closely related to its colloquial usage.

Suppose I make N measurements of some quantity T, with resulting values $T_1, T_2, \ldots, T_N$. Then the average, or *mean*, of these measurements is simply:

$$\langle T \rangle = \frac{1}{N} \sum_{j=1}^{N} T_j, \tag{3.3}$$

where I have used the common notation $\langle T \rangle$ to denote the average of T.

I could also calculate the *theoretical* average value of T—the value that I would *expect* to obtain *if* I made (very) many measurements. Suppose that a (single) measurement of T must yield one of a set of β possible values. If I know the probabilities for obtaining each of these β values, then the expected average value of T is

$$\langle T \rangle = \sum_{k=1}^{\beta} Prob(T_k)T_k. \tag{3.4}$$

It may not be obvious why Eq. (3.4) should hold, or how it's connected to Eq. (3.3). Equations (3.3) and (3.4) differ because Eq. (3.4) is the ideal, expected average in terms of theoretical quantities (probabilities), while Eq. (3.3) is the average of a finite set of N actual measurements. Suppose the value $T_j = \alpha$ occurs N_α times in our N measurements. Each time it occurs, a term of the form α/N appears in the sum in Eq. (3.3). The sum of all such terms is: $(N_\alpha/N)\alpha$. In the limit $N \to \infty$, we have $(N_\alpha/N)\alpha \to Prob(\alpha)\alpha$ (cf. Eq. (3.1)). If we do this for all T values, then Eq. (3.3) becomes Eq. (3.4).

If T is a continuous variable, we have a probability distribution (or density), denoted $\rho(T)$, with dimensions *probability*/T. The sum becomes an integral, the index k becomes superfluous, and Eq. (3.4) takes the form

$$\langle T \rangle = \int_{T_{min}}^{T_{max}} \rho(T)T \, dT. \tag{3.5}$$

Here T_{min} and T_{max} are the minimum and maximum values that T may assume.

In quantum mechanics, an ideal, expected average value—such as $\langle T \rangle$ in Eq. (3.4) or (3.5)—is usually referred to as an *expectation value*. Thus, the expectation value is a *calculated* (rather than measured) average. The term expectation value is slightly misleading because, despite its name, we do not, in general, actually expect to obtain the expectation value in any given measurement. For example, if the only two possible results of a measurement are $+1$ and -1, and if each result is equally probable, then the expectation value is zero. But we will *never* obtain this expectation value in any measurement. (Similarly, in any given roll of the die in Sec. 3.1.1, we will never obtain the expectation value for the number of dots. Calculate it!)

For discrete eigenvalues, the quantum expectation value corresponds to Eq. (3.4). Given, for example, the eigenvalue equation $\hat{B}\phi_j = b_j\phi_j$, and the state $\Omega = \sum_i \alpha_i\phi_i$ (from Sec. 2.3.1), the expectation value $\langle B \rangle$ is of the form

$$\langle B \rangle = \sum_i Prob(b_i)b_i = \sum_i |\alpha_i|^2 b_i. \tag{3.6}$$

In the context of wavefunctions—which is how students often first encounter quantum-mechanical expectation values—we have

$$\langle A \rangle = \int_a^b \Psi(x)^* \hat{A} \Psi(x) dx, \qquad (3.7)$$

where the observable is A (corresponding to the operator $\hat{A}$), and the system is in the state $\Psi(x)$. Here the interval (a, b) is the region within which $\Psi(x)$ may be non-zero (and thus, within which the particle may be found upon measurement). Clearly, the limits of integration could be extended to $(-\infty, \infty)$ without affecting the integral.

Consider $\langle x \rangle$, the expectation value for position. In this case it's easy to see that $\langle x \rangle$ takes the form of Eq. (3.5). From Section 2.3.2, we know that if $\hat{x}$ acts on $\Psi(x)$, then $\hat{x} = x$. Thus, for $\langle x \rangle$, Eq. (3.7) may be written as

$$\langle x \rangle = \int_{-\infty}^{\infty} \Psi(x)^* x \Psi(x) dx = \int_{-\infty}^{\infty} \Psi(x)^* \Psi(x) x dx = \int_{-\infty}^{\infty} |\Psi(x)|^2 x dx.$$
$$(3.8)$$

We know from Section 2.3.2 that $|\Psi(x)|^2$ is a probability density, as is $\rho(T)$ in Eq. (3.5). So there's a straightforward connection between Eq. (3.5)—a result from probability theory—and the quantum expectation value $\langle x \rangle$. The crucial step in Eq. (3.8) was taking $x\Psi(x) = \Psi(x)x$—a mathematically trivial operation, but *only* because $\hat{x} = x$. It was this step that led to the connection with Eq. (3.5), through the analogous roles of $|\Psi(x)|^2$ and $\rho(T)$.

If we consider Eq. (3.7) for some $\hat{A}$ that—unlike $\hat{x}$ in Eq. (3.8)—must be treated as an operator, rather than as a variable, it's no longer obvious how to interpret $\Psi(x)^* \hat{A} \Psi(x)$. The simple link to Eq. (3.5) is thus lost. Even in such cases, however, Eq. (3.7) still holds.[4]

3.1.3 Uncertainties

You'll hear much about *uncertainty* in quantum mechanics. But like probabilities and averages, the concept of uncertainty is fundamentally statistical, not quantum-mechanical.[5]

The expectation value (i.e., the average value) provides important information about a probability distribution—but there's much that it doesn't tell us. Consider two possible probability distributions, denoted a and b, of some variable: $Prob_a(+1) = Prob_a(-1) = 1/2$, and $Prob_b(+5) = Prob_b(-5) = 1/2$. (See Fig. 3.2.) It's easy to see that although the distributions share the same expectation value ($\langle a \rangle = \langle b \rangle = 0$), they are *much*

---

[4] See, for example, Bohm (1951), Sections 9.5-9.6; Schiff (1968), pp. 27-28. See also Section 4.4.3.

[5] My discussion of uncertainties here will be minimal; more detailed treatments are readily available. See, for example, Taylor (1997).

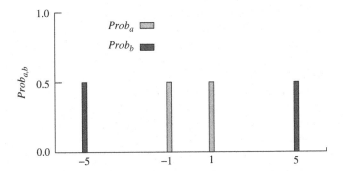

Fig. 3.2 Two non-identical probability distributions with identical expectation values ($\langle a \rangle = \langle b \rangle = 0$).

different, as is evident from their different "spreads". Clearly, we could easily generate an unlimited number of widely varying distributions, all with expectation value zero.

Uncertainty provides a means of characterizing the spread of a distribution. The most straightforward (and conventional) way to do this is by defining the uncertainty of a distribution to be its standard deviation, σ. For a set of N measurements of a quantity T, the standard deviation is

$$\sigma(T) = \sqrt{\frac{\sum_{j=1}^{N}(T_j - \langle T \rangle)^2}{N}}. \tag{3.9}$$

Evidently, $\sigma(T)$ is the *root* of the *mean* of the *squares* of the *deviations* from $\langle T \rangle$; the standard deviation is thus also called the *root mean square deviation*.

To calculate uncertainties theoretically, we must cast the standard deviation into probabilistic language. Take T_k to represent the set of all M theoretically possible measurement results. Then $\left\langle (T_k - \langle T \rangle)^2 \right\rangle$ is the mean of the square of the deviations of the T_ks from $\langle T \rangle$, and Eq. (3.9) becomes

$$\sigma(T) = \sqrt{\left\langle (T_k - \langle T \rangle)^2 \right\rangle} = \sqrt{\frac{\sum_{k=1}^{M} \left(T_k^2 - 2T_k\langle T \rangle\right) + M\langle T \rangle^2}{M}}$$

$$= \sqrt{\langle T^2 \rangle - \langle T \rangle^2}. \tag{3.10}$$

If you use Eq. (3.10) to calculate the uncertainties for the two distributions in Fig. (3.2), you can see that σ does indeed characterize their spreads (try it!). But these distributions are obviously very simple—so what does the uncertainty tell us about the spread of an arbitrary distribution?

Use of the term "uncertainty" in an experimental context usually refers to an inability to make exact measurements. This inability manifests itself in the spreading out of a set of measurements around an average measured value. Typically, the distribution of measurement results approaches the form of a Gaussian, or *normal,* distribution (as in Fig. 3.1(*b*)). You may have learned elsewhere that the standard deviation (the uncertainty) represents the 68% "confidence level"—that we expect 68% of our measurements will lie within one standard deviation of the average. This is correct, *if* the distribution is indeed Gaussian, as we expect for most experimental uncertainties. But we really can't say much about the meaning of the uncertainty for an arbitrary distribution except that, roughly speaking, a larger uncertainty means a more spread-out distribution.

Because quantum states (and their corresponding probability distributions) generally are *not* Gaussians, we can't easily characterize the degree of spreading represented by the uncertainty in quantum mechanics. Nevertheless, uncertainties remain useful in quantum mechanics, partly because it is often the relation between the uncertainties in different distributions, rather than the uncertainty of any individual distribution, that is of interest.

Probabilities, averages, and uncertainties—these are all the statistical concepts needed to develop basic quantum mechanics, and I again emphasize that they are fundamentally statistical in nature, not quantum-mechanical. Having introduced these basic statistical concepts, I now introduce the statistical interpretation of quantum mechanics.

3.2   The Statistical Interpretation

Most texts adopt—implicitly or explicitly—the so-called *Copenhagen interpretation* of quantum mechanics. Like other, conflicting interpretations, Copenhagen is an attempt to tell us what's really "going on"—to inform us of the quantum world beyond the formalism itself.

By contrast, the statistical interpretation arguably is not an *interpretation,* but a broad *framework* that describes how quantum mechanics works in actual practice. You may worry that by learning quantum mechanics from this perspective, you'll be at a disadvantage. On the contrary, the statistical interpretation provides an understanding one can have confidence in: because it's "just" a framework, it remains compatible with other approaches, such as Copenhagen, while avoiding many conceptual puzzles that arise within them.

But what *is* the statistical interpretation? In a sense, it simply amounts to the following edict: take seriously what quantum mechanics *does* tell us, and don't take seriously what quantum mechanics *doesn't* tell us.

So, what does quantum mechanics tell us? First, it provides a prescription for calculating the quantum states associated with a physical system, including how these states evolve in time.[6] And, if we can calculate the quantum state that properly describes a particular system, quantum mechanics tells us how to obtain the probability distributions for the results of measurements performed on that system.

That last paragraph could stand lots of unpacking. For now, however, I only wish to make a simple but crucial point. I will do so with a conceptual example that is essentially devoid of mathematics. Consider a system with only two possible energies, E_1 and E_2. In general there is an infinite ensemble (set) of quantum states that *could* describe the system. However, for a particular physical situation, the system must be described by one particular quantum state,[7] which I'll denote Ψ. Suppose now that we have correctly calculated the quantum-mechanical state Ψ, which then yields the following probability distribution for energy measurements on the system: $Prob(E_1) = \frac{1}{3}$, and $Prob(E_2) = \frac{2}{3}$. So far, we've simply followed the quantum prescription:

- calculate a quantum state that correctly describes the system of interest,
- use the quantum rules to obtain the probability distribution.

But what does this probability distribution mean in a real, practical sense?

Imagine we have an ideal energy measuring device in our laboratory. Quantum mechanics never predicts that we will obtain some intermediate value in any given energy measurement (see Section 2.3). Thus, the energy probability distribution becomes manifest *not* in a single measurement, but only in a distribution built up from an ensemble of measurements. Just as we *never* obtain a value between heads and tails in an individual coin toss (assuming that the coin never lands balanced on its edge), we *never* obtain a value between E_1 and E_2 (such as, say, $\frac{1}{3}E_1 + \frac{2}{3}E_2$) on any individual energy measurement: we always obtain *either* E_1 *or* E_2. As the number of measurements, N, increases, we obtain E_1 in approximately $\frac{1}{3}$ of our measurements, and E_2 in approximately $\frac{2}{3}$ of our measurements. And as $N \to \infty$, these approximations become better. All of this is precisely what we expect on the basis of our discussion of *non*-quantum probabilities in Section 3.1.1.

In the laboratory, then, we generally must make many measurements before the probability distribution becomes manifest. And although our example dealt with energy, this holds regardless of what quantum observable one measures—energy, momentum, angular momentum, or position. Moreover, it holds regardless of interpretation. Copenhagenist or not, very

---

[6] This is not to say that the prescription is easily filled. One may face formidable challenges, both in terms of determining the correct Hamiltonian to use, and in terms of carrying out the prescribed calculations, that is, solving for the quantum state.

[7] If we considered mixtures, this statement would require modification (see Section 2.3.1).

few physicists would disagree with this "many measurements" view of experiments.

Now it's one thing to "know" something, and often quite another to intellectually digest it until it becomes a mode of thought. Perhaps what I've said seems simple enough: Treat the quantum probabilities like non-quantum probabilities, in that they become manifest through repeated measurements. But this point is so fundamental that it's worth restating.

> The fundamental role of quantum mechanics is to provide a means to calculate possible measured values, and the probability distributions corresponding to those values, for measurements performed on physical systems. These probabilities become evident only in an ensemble of measurements, *never* in a single measurement.

And just as probabilities become manifest in an ensemble of measurements, so do averages and uncertainties. In the laboratory, these quantities arise through application of Eqs. (3.3) and (3.9) to an ensemble of measurements, *never* to a single measurement.

Note carefully, however, that quantum mechanics tells us nothing about measurable quantities *absent* in a measurement. Take our example: quantum mechanics proper simply does not address questions such as, What is the system's energy between measurements? Is the concept of energy undefined except when measured? Why, when measured, is the energy always E_1 or E_2?

Such questions are deeply interesting and important. For example, if we impose completeness (see Section 2.1), then the quantum state must somehow contain within it everything that can, *in principle*, be known about the system. Thus, the state itself must somehow provide answers to our questions—yet none are forthcoming. In the statistical interpretation, we choose to simply ignore such questions. And in the end they remain unresolved, for our instruction set—the quantum formalism—provides no answers. None are needed, however, to pursue our desideratum: developing a clear and coherent means to think about quantum mechanics.

In the statistical interpretation, then, we generally refrain from looking beyond quantum mechanics' operational role as a means of calculating possible measured values and the corresponding probability distributions. And these quantum probabilities become manifest in the same manner as non-quantum probabilities: in an ensemble of measurements of the quantity of interest.

3.3 Bohr, Einstein, and Hidden Variables

3.3.1 Background

The interpretation of quantum mechanics has been a contentious issue in physics for decades. In older books one typically finds a matter-of-fact acceptance of the Copenhagen interpretation. But in recent years interest in other interpretations of quantum mechanics has enjoyed a renaissance, and researchers in the field generally believe that the correct interpretation remains an open question. This has, in some measure, provided the impetus for my approach: regardless of any unresolved metaphysical questions, we can agree on the *operational* meaning of quantum mechanics as embodied in the statistical interpretation.

I will not discuss the historical, philosophical, and technical aspects of the interpretation of quantum mechanics in any depth—to do so could easily fill a book in itself. I hope only to provide some background, and to dispel some persistent myths.

The formalism of non-relativistic quantum mechanics was completed during 1925 and 1926. Also in 1926, Max Born showed how to obtain probability distributions from the quantum formalism. This essentially completed what I have called the statistical interpretation of quantum mechanics. Operationally, this is sufficient, but it doesn't tell us what's really going on—what the quantum formalism, particularly the quantum state, means *physically*. And it's the task of attaching physical meaning to the mathematical formalism of quantum mechanics that has been the focus of so much debate.

The 1927 Solvay Conference, in Brussels—a meeting of the great physics luminaries of the time—marks a turning point. There, Niels Bohr and Albert Einstein engaged in a famous, extended debate about the interpretation of quantum mechanics. Bohr espoused what would soon become the almost universally accepted view, Einstein vigorously calling that same view into question.

The Bohr–Einstein debate is often portrayed as a decisive defeat for Einstein—an aging physicist who simply could not abandon his antiquated philosophical commitment to causality, and thus stubbornly opposed quantum mechanics until his death nearly thirty years later. But history, including history of science, is often not so simple as its folklore rendition, and this portrayal of the debate is open to serious questions.

Nevertheless, the belief that Einstein had been defeated was widespread, and this contributed to the rapid rise to dominance of Bohr's ideas, and what came to be known as the Copenhagen interpretation. After the Solvay Conference, very few physicists seriously questioned Copenhagen. Those who did were marginalized, largely being viewed as the old guard, incapable of effecting the revolution in thinking required by quantum mechanics.

There are various versions of the Copenhagen interpretation.[8] But an insistence on *completeness*, which implies that the quantum state contains

---

[8] So, just what *is* the Copenhagen interpretation? This question, including historical background, is discussed from disparate perspectives in Cushing (1994), Chapter 3, and Howard (2004).

all information about the system it describes, is common to all of them. Even in 1927 completeness was gaining acceptance:

> We maintain that quantum mechanics is a complete theory; its basic physical and mathematical hypotheses are not further susceptible of modifications.[9]
> (Werner Heisenberg and Max Born)

This would be a remarkably forceful assertion to make about even a mature scientific theory. Heisenberg and Born did so when quantum mechanics, as a coherent theory, was only some two years old.

Completeness is clearly incompatible with the existence of any information other than that contained in quantum mechanics itself. Thus, an insistence on completeness rules out the existence of *hidden variables* that contain information beyond that embodied in the quantum state. The precise nature of such variables is, for our discussion, unimportant. The point is that because hidden variables theories are inconsistent with completeness, the existence of hidden variables would invalidate the Copenhagen interpretation.

3.3.2 Fundamental Issues

Here I briefly discuss what are likely the three most important topics in foundations of quantum mechanics. The first two are problems, in the sense that no consensus has been reached as to how, or even whether, they will be resolved. They serve to illustrate that major interpretational problems remain in quantum mechanics. The last topic, nonlocality, is not so much a problem as a profound result. I introduce nonlocality here because of its fundamental importance, but also to warn against reading *too* much into this remarkable result.

The Measurement Problem. The measurement problem, which arises from Postulate 3 (Section 2.3), is one of the most famous and stubborn in quantum mechanics. Often it is characterized as "the collapse of the wave function," indicating that when a measurement is performed, the result will be one of the eigenvalues of the measured observable. But this description fails to convey the depth of the difficulty. That the wave function (quantum state) collapses when a measurement is performed is indeed surprising, but it is not obviously inconsistent. The real crux of the measurement problem is not simply *that* collapse occurs, but rather *how* it occurs.

It can be shown (without undue difficulty) that collapse cannot occur due to "regular" time evolution by the Schrödinger equation (as in Postulate 1). That is, we must suspend the Schrödinger equation during a

---

[9] From a lecture delivered at the 1927 Solvay Conference; quoted in Jammer (1974), p. 114.

measurement, so as to effect the collapse. But must not a measurement simply be a particular kind of physical interaction—different in details, but not in kind, from non-measurement interactions? What, then, can justify "turning off" the Schrödinger equation only during a measurement? As John Bell asked, "are we not obliged to admit that more or less 'measurement-like' processes are going on more or less all the time more or less everywhere? Is there ever then a moment when there is no [collapsing] and the Schrödinger equation applies?"[10] This, more so than collapse *per se*, lies at the heart of the measurement problem.

The Classical Limit. Special relativity is a more fundamental theory than Galilean relativity, so that Galilean relativity should emerge from special relativity in some appropriate limit. And indeed, this is just what happens: for velocities much smaller than light speed, special relativity reduces to Galilean relativity.[11] *The classical limit of quantum mechanics* refers to the belief that quantum mechanics is a more fundamental theory than classical mechanics, so that classical mechanics should emerge from quantum mechanics in some appropriate limit.

Ideally, this emergence would result from some simple limiting procedure, as for relativity. Alas, no generally applicable route to the classical limit has ever been found. Despite what some textbooks suggest, it is not clear how, or even if, the classical limit can be realized. It now appears that any resolution of the classical limit problem will require, at the least, *both* a limiting procedure *and* accounting for the interaction of quantum systems with the environment, through a sophisticated approach known as *decoherence.*

Non-locality. If a theory is *local,* there are no superluminal interactions (i.e. interactions that travel at faster than light speed) between spatially separated systems.[12] In 1964 John Bell published his renowned namesake theorem, which showed that any hidden-variables theory that obeys locality must make (some) predictions that are inconsistent with those of quantum mechanics. Experimental tests confirmed quantum mechanics, effectively ruling out local hidden-variables theories. But even though local hidden-variables theories conflict with experiment, couldn't *non*-local hidden-variables theories still be valid?

To address this question, an argument like the following is sometimes advanced. In the wake of experimental tests of Bell's theorem, the choice is between standard (Copenhagen), local quantum mechanics, and non-local hidden-variables theories. But because non-locality violates special relativity (i.e. no signals may travel faster than light speed), which has been convincingly confirmed, non-local hidden-variables theories are unacceptable. Thus, *all* hidden-variables theories fail, local *or* non-local.

---

[10] Bell (1987), p. 117.
[11] Actually, this oft-repeated statement isn't *quite* true; see Baierlein (2006).
[12] Some authors define locality somewhat differently.

This argument is problematic. The choice between a local and non-local theory is illusory—because standard quantum mechanics, also, is non-local. Says Tim Maudlin, "Stochastic [Copenhagen-type] theories involve an *obvious* element of superluminal causation. Deterministic [hidden-variables] theories were the *only hope* for locality from the beginning, a hope Bell extinguished [italics added]."[13] Or as James Cushing put it, "One of the central lessons we may draw from Bell's theorem and from the analysis resulting from it is that such nonlocality appears to be a feature of our *world,* not just of this or that *theory* of physical phenomena. That being the case, nonlocality itself gives us little reason to choose Copenhagen over [hidden variables]."[14] Bell's theorem is simply mute in the debate over the validity of non-local hidden-variables theories—the correct interpretation of quantum mechanics remains an open question.

Even though it cannot resolve the hidden-variables debate, Bell's theorem remains of great interest—non-locality *is* a profoundly important discovery. "For those interested in the fundamental structure of the physical world," writes Maudlin, "the experimental verification of violations of Bell's inequality constitutes the most significant event of the last half-century."[15]

3.3.3 Einstein Revisited

Because of his opposition to the Copenhagen interpretation, Einstein was widely regarded as out of touch with the new physics, unable to adapt to the bold ideas of quantum mechanics. You may hear that Einstein could never accept quantum mechanics, despite its overwhelming success, and despite the compelling arguments in its favor, such as those advanced by Bohr at the 1927 Solvay Conference.

But Einstein *did* accept the predictive validity of quantum mechanics—that it provides a correct description of the world. In essence, he accepted the statistical interpretation. What he did not accept, what his objections were rooted in, was an insistence on quantum mechanics' *completeness.*

Regarding the fundamental role of probability in quantum mechanics, Einstein famously stated that "God does not play dice"—to which Bohr reportedly responded: "Stop telling God what to do."[16] And many still seem to regard Einstein's remark as a self-inflicted wound: an indication of his stubborn unwillingness to change, his lamentably archaic attitude in the face of the new physics. But Einstein's insistence that chance *cannot* be a fundamental principle of nature was no more dogmatic than the insistence of so many of his contemporaries that it *must* be. Perhaps a worthy retort to Bohr would have been: "Stop telling God he *must* play dice."

[13] Maudlin (1994), pp. 138–139.
[14] Cushing (1994), p. 47.
[15] Maudlin (1994), p. 4.
[16] Calaprice (2000), pp. 251–252.

Evidently Einstein did object to the notion that chance operates at the most fundamental level of nature. But a far deeper objection was not to chance, but to physical *unreality*, which, Einstein argued, arises from the insistence on completeness.[17]

The attitudes of Einstein, and those few like him, towards quantum mechanics are often seen as archaic and outmoded. But one could also adopt a sympathetic perspective. Facing a tide of scientific opinion which had turned decidedly against them, these brave souls persevered. And Einstein, in particular, seemed unperturbed by it all. Near the end of his life, he wrote that...

> Few people are capable of expressing with equanimity opinions which differ from the prejudices of their social environment. Most people are even incapable of forming such opinions.

This indictment was leveled at the community at large, but might it also have been directed at the community of scientists? Einstein's words form a fitting if unintentional tribute to their author—a man who never relinquished his intellectual integrity or independence, despite the prejudices of his scientific environment; a man who grew estranged from the scientific community, but never from science.[18]

3.4 Problems

1. Suppose the possible measured values of r for some quantum system are:

$$r_j = \frac{k}{j^2}(-1)^j,$$

where k is the index that appears in the c_ks, the expansion coefficients. You have an excellent lab at your disposal, operated by highly competent experimental physicists.

(i) As a preliminary test, the lab prepares systems with the following expansion coefficients:

$$c_1 = \frac{1}{\sqrt{15}}, \quad c_2 = \sqrt{\frac{5}{15}}, \quad c_5 = \frac{3i}{\sqrt{15}}.$$

Calculate $\langle r \rangle$, the expectation value of the quantity r, in the above state.

(ii) Your experimentalist colleagues prepare 75 systems in the above state, and measure r on each of them. They obtain r_1 in 6 runs, r_2 in

---

[17] See Ballentine (1972). A fascinating correspondence amongst Einstein, Max Born, and Wolfgang Pauli sheds light on this topic. See: Born (1971), pp. 205–228. Note that "statistical interpretation" is used therein in a much different sense than I have used it.

[18] Quoted in Einstein (1984), p. 102.

22 runs, and r_5 in 47 runs. Calculate the experimental average, and compare it to $\langle r \rangle$ above.

(iii) In the actual experiments you wish to run, the situation is not so simple. Now there may be *many* expansion coefficients. Moreover, you cannot afford to make simple calculational errors. Write a program (using Fortran, MATLAB, etc.) to calculate $\langle r \rangle$ from a known set of coefficients (which you enter into the program).

(iv) Your experimentalist colleagues prepare the systems with the following coefficients.

$$c_1 = \sqrt{\tfrac{1}{50}} \quad c_4 = \sqrt{\tfrac{5}{50}} \quad c_7 = -i\sqrt{\tfrac{2}{50}} \quad c_{10} = \sqrt{\tfrac{1}{50}} \quad c_{13} = -i\sqrt{\tfrac{5}{50}}$$

$$c_2 = -\sqrt{\tfrac{3}{50}} \quad c_5 = \sqrt{\tfrac{1}{50}} \quad c_8 = \sqrt{\tfrac{6}{50}} \quad c_{11} = i\sqrt{\tfrac{3}{50}} \quad c_{14} = \sqrt{\tfrac{8}{50}}$$

$$c_3 = \sqrt{\tfrac{2}{50}} \quad c_6 = \sqrt{\tfrac{3}{50}} \quad c_9 = \sqrt{\tfrac{2}{50}} \quad c_{12} = \sqrt{\tfrac{7}{50}} \quad c_{15} = \sqrt{\tfrac{1}{50}}$$

Now run your program with these coefficients. (You can use Part (i), above, to check it.) It's easy to see the power (and accuracy, *if* you write your program correctly!) of using a computer to calculate such things, rather than calculating "by hand".

2. Consider a normal, six-faced die. Each face has from one to six dots on it (each number of dots appearing only once). On any die roll, each face has an equal chance of coming up.

(i) Calculate $< dots >$, the expected (theoretical) average number of dots.

(ii) Suppose you roll the six-faced die N times, and obtain the following results.

No. of dots	No. of times obtained
One	$N/6$
Two	$N(3/24)$
Three	$N(3/16)$
Four	$N/6$
Five	$N(7/48)$
Six	$N(5/24)$

Calculate $< dots >$, the average measured number of dots. Should this quantity be called an expectation value? How about $< dots >$ from Part (i)? Should the quantities in the right-hand column above properly be called probabilities?

(iii) Calculate the uncertainty for the measurement results of Part (ii).

(iv) Calculate the theoretical uncertainty for the system of Part (ii). Does your result seem reasonable? Does it agree with the result of Part (iii)? Why, or why not?

(v) The *Cosa Nostra* casino in Las Vegas uses "loaded" dice. The probability distribution for such a die is:

No. of dots	Probability
One	5/24
Two	1/6
Three	1/8
Four	1/8
Five	1/6
Six	5/24

Calculate $< dots >$ for such a die. Also calculate the uncertainty.

(vi) Plot the probability distribution for the die of Parts (i–iv). Show the uncertainty, also. Then do the same for the die of Part (v).

4

The Structure of Quantum States

Before I had studied Zen for thirty years, I saw mountains as mountains and waters as waters. When I arrived at a more intimate knowledge, I came to the point where I saw that mountains are not mountains, and waters are not waters. But now that I have got its very substance I am at rest. For it's just that I see mountains once again as mountains, and waters once again as waters.
Ch'ing-Yuan[1]

Quantum mechanics may at first seem deceptively simple. But as its detailed structure is unmasked, it becomes complex and mystifying. In time, an inner simplicity is revealed—one arising now not from ignorance, but from insight into quantum mechanics' fundamental structure.

Perhaps you now feel that you've acquired a good grasp of the concept of a quantum state. But grasping the essence of quantum mechanics requires a deep understanding of the state, and the simple picture presented so far is insufficient to develop such an understanding.

This chapter first provides the requisite mathematical background, and then introduces Dirac notation—an immensely useful and elegant language with which to represent states. The power of Dirac notation begins to reveal itself as the *scalar product*—a pervasive concept in quantum mechanics—is developed. Finally, the key quantum concept of a *representation* is explored. Only when the relations amongst eigenstates, superpositions, and representations are understood can the fundamental structure of quantum states be revealed.

4.1 Mathematical Preliminaries

4.1.1 Vector Spaces

Our goal is the development of a working understanding of quantum mechanics, rather than a rigorous mathematical exposition. To better understand quantum states, however, we'll need some fundamental mathematical concepts—these I present in this section and the next.

---

[1] Watts (1989), p. 126.

Although elementary vectors such as force and momentum are, at most, three-dimensional, the mathematical concept of a vector may be extended to any (integer) number of dimensions. The familiar scalar (dot) product of two Cartesian vectors is immediately generalized to n dimensions—the scalar product of two n-dimensional vectors $\vec{A}$ and $\vec{B}$ is

$$\vec{A} \cdot \vec{B} = \sum_{k=1}^{n} A_k B_k = A_1 B_1 + A_2 B_2 + \cdots + A_n B_n. \tag{4.1}$$

The geometric interpretation of $\vec{A} \cdot \vec{B}$ is that it returns a scalar, the size of which is the length of the *projection* of vector $\vec{A}$ onto vector $\vec{B}$, multiplied by the length of vector $\vec{B}$. (We could equally well think of $\vec{A} \cdot \vec{B}$ as giving the length of the projection of $\vec{B}$ onto $\vec{A}$, multiplied by the length of $\vec{A}$.)

If we insist that $\vec{A}$ and $\vec{B}$ be *unit* vectors, then the interpretation is simplified. Since the length of each vector is 1, $\vec{A} \cdot \vec{B}$ is simply the projection of either one of the vectors onto the other. In the xy plane, for example, if $\vec{A}$ is the unit vector in the x-direction, and $\vec{B}$ a unit vector at an angle of $\pi/3$, then $\vec{A} \cdot \vec{B} = \cos(\pi/3) = 1/2$, which is the projection of $\vec{A}$ onto $\vec{B}$.

Imagine a Cartesian coordinate system in a three-dimensional space. Each of the three unit (length 1) vectors $\hat{e}_1$, $\hat{e}_2$, and $\hat{e}_3$ lies along one of the three Cartesian axes.[2] These unit vectors are *orthonormal*: the scalar (or dot) product of any one of the unit vectors with any other one is 0 (orthogonality), and the scalar product of any one of the unit vectors with itself is 1 (normality, i.e., the vectors are of unit length).

Clearly, *any* vector, say $\vec{V}$, in this three-dimensional space may be written as a *linear combination* (a weighted sum) of $\hat{e}_1$, $\hat{e}_2$, and $\hat{e}_3$:

$$\vec{V} = \alpha \hat{e}_1 + \beta \hat{e}_2 + \gamma \hat{e}_3 \tag{4.2}$$

where α, β, and γ are suitably chosen constants. We say that $\hat{e}_1$, $\hat{e}_2$, and $\hat{e}_3$ *span* the space, or that they constitute a *complete set* of vectors, or a set of *basis vectors*, in the space.

However, we could write $\vec{V}$ as a linear combination of *any* complete set of vectors in our space. For example, we could define a new Cartesian coordinate system by rotating our original system. Our unit vectors along the new Cartesian axes then point in different directions than $\hat{e}_1$, $\hat{e}_2$, and $\hat{e}_3$, but they still constitute a complete set (and are still orthornormal), so we can write $\vec{V}$, or any other vector in the space, as a linear combination of these new basis vectors.

In an n-dimensional space, any set of n orthogonal vectors forms a complete set. However, non-orthogonal vectors may also form complete sets. The necessary condition such that a set of n vectors forms a complete

---

[2] Note that here the "hat" notation denotes a unit vector, not an operator.

set—such that any vector in the n-dimensional space may be written as a linear combination of the vectors in the set—is that they be *linearly independent*. Linear independence means, in essence, that none of the n vectors in the set can be written as a linear combination of the other $n-1$ vectors in the set.

As a simple example, consider the xy plane ($n = 2$). The Cartesian unit vectors $\hat{x}$ and $\hat{y}$ are orthonormal (their scalar product fulfills the orthonormality conditions set forth above), and thus linearly independent: we cannot write $\hat{x}$ as a linear combination of the (only other) vector $\hat{y}$ (and vice versa). That is, there is no constant α such that $\hat{x} = \alpha\hat{y}$. Accordingly (and obviously), any vector in the xy plane may be written as a linear combination of the set $\hat{x}$ and $\hat{y}$.

Now, however, consider the set of unit vectors: $\hat{x}, (\hat{x}+\hat{y})/\sqrt{2}$. It is impossible to write either vector as a linear combination of the other, so these two vectors *are* linearly independent. They are *not*, however, orthonormal (or even orthogonal). A bit of thought should convince you, though, that any vector in the xy plane may be written as a linear combination of $\hat{x}$ and $(\hat{x} + \hat{y})/\sqrt{2}$.

Orthogonality implies linear independence, but linear independence does *not* imply orthogonality. Orthonormality of a set of n vectors in an n-dimensional space is thus a sufficient, but not necessary, condition for the vectors to span the space. Nevertheless, it is usually convenient to use orthonormal basis vectors, just as we would usually choose $\hat{x}$ and $\hat{y}$, not $\hat{x}$ and $(\hat{x} + \hat{y})/\sqrt{2}$, as a basis set in the xy plane.

So if $\vec{N}$ is a vector in an n-dimensional space (where n could be infinite), and $\{\hat{v}_i\}$ is a set of n linearly independent basis vectors in the space, then $\vec{N}$ may be written as the linear combination,

$$\vec{N} = \sum_i c_i\hat{v}_i = c_1\hat{v}_1 + c_2\hat{v}_2 + \cdots + c_n\hat{v}_n, \tag{4.3}$$

where the c_is are called *expansion coefficients*. But how do we find the c_is? For the general case, where our only constraint on the basis vectors is that they are linearly independent, finding the c_is may take some effort. If, however, the basis vectors are orthonormal, finding the c_is becomes straightforward.

If $\vec{a}$ is some vector in the xy plane, and our orthonormal basis set is $\hat{x}$ and $\hat{y}$, then $\vec{a}$ can be written as

$$\vec{a} = c_1\hat{x} + c_2\hat{y} = a_x\hat{x} + a_y\hat{y} = (\vec{a} \cdot \hat{x})\hat{x} + (\vec{a} \cdot \hat{y})\hat{y}. \tag{4.4}$$

That is, c_1 and c_2 are simply a_x and a_y, the (magnitudes of the) projections of $\vec{a}$ onto the x and y axes, respectively (Fig. 4.1); a_x and a_y are, in turn, just the scalar products of $\vec{a}$ with the unit vectors $\hat{x}$ and $\hat{y}$, respectively (cf. Eq. (4.1)). This result immediately generalizes to n dimensions. For

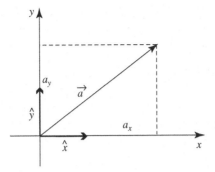

Fig. 4.1 Projection of the vector $\vec{a}$ onto the $\hat{x}$ and $\hat{y}$ basis vectors.

example, if the set $\{\hat{v}_i\}$ is not just linearly independent, but orthonormal, $\vec{N}$ may be written as

$$\vec{N} = \sum_i c_i \hat{v}_i = \sum_i (\vec{N} \cdot \hat{v}_i)\hat{v}_i. \tag{4.5}$$

4.1.2 Function Spaces

Now I introduce another, quite abstract, sort of mathematical "space": *function space*. We may visualize vectors in space—even in an abstract, n-dimensional space. But this visualizability evaporates in function space, and that can lead to confusion. The problem partly stems from the fact that, in common usage, the word "space" suggests something that *should* be visualizable. In mathematics, however, the word space refers not to real space, or even to some analog thereof, but to a set of rules that govern the behavior of specified mathematical objects. Thus, to say that an object is in function space simply means that it behaves according to certain rules.

The bad news, then, is that function spaces are much more abstract than the "regular" vector spaces we've encountered so far. The good news is that a close analogy exists between vectors in a vector space, and mathematical functions in a function space. By importing the concepts of vector spaces into function space, the abstract nature of the latter becomes more manageable.

Suppose two vectors in an n-dimensional vector space, $\vec{C}$ and $\vec{D}$, are orthonormal. The orthonormality condition may be expressed as

$$\vec{C} \cdot \vec{D} = \sum_{i=1}^{n} C_i D_i = C_1 D_1 + C_2 D_2 + \cdots + C_n D_n = \left\{ \begin{array}{ll} 0, & \text{if } \vec{C} \neq \vec{D} \\ 1, & \text{if } \vec{C} = \vec{D} \end{array} \right\}.$$

$$\tag{4.6}$$

We know that any set of n such orthonormal vectors forms a complete set, and any vector in the space may be written as a linear combination of the basis vectors that comprise such a set.

Function spaces are defined in analogy to vector spaces. Just as a class of vectors forms a vector space, so does a class of functions form a function space. For example, the set of second-order polynomials on the interval $(0, 1)$ forms a function space. Another example is the set of real functions of x that are square-integrable on some interval (x_1, x_2).[3]

For complex functions $g_j(x)$ and $g_k(x)$, the orthonormality condition is

$$\int_{x_1}^{x_2} g_j(x)^* g_k(x) dx = \delta_{jk} = \left\{ \begin{array}{ll} 0, & \text{if } j \neq k \\ 1, & \text{if } j = k \end{array} \right\}, \qquad (4.7)$$

where δ_{jk}, the *Kronecker delta*, is defined as indicated by the right-hand side of Eq. (4.7). Equations (4.6) and (4.7) are clearly analogs of each other. For vectors, we take the products of the corresponding components of each vector, and sum over all such products. For functions, we take the product of the two functions' values at each x, and integrate over x.[4]

In analogy with vectors, any function in some space can be written, or expanded, as a linear combination of any complete set of *basis functions* for that space. In an n-dimensional function space, any set of n linearly independent functions forms a complete set. And, as for vectors, orthonormality is a stronger condition than linear independence, so any set of n orthonormal functions constitutes a complete set.

Explicitly, if $\{g_m(x)\}$ constitutes our complete set of basis functions, then,

$$f(x) = \sum_n a_n g_n(x), \qquad (4.8)$$

where the a_ns are expansion coefficients. If the $g_n(x)$s are orthonormal, the a_ns take a form analogous to that for the expansion coefficients for vectors in an orthonormal set:

$$a_m = \int_{x_1}^{x_2} g_m(x) f(x) dx. \qquad (4.9)$$

In Eq. (4.5), the expansion coefficients take the form $\vec{N} \cdot \hat{v}_i$, that is, the projection of the arbitrary vector $\vec{N}$ onto each of the basis vectors. In Eq. (4.9), the expansion coefficients take the form of the projection of the arbitrary function $f(x)$ onto each of the basis functions $g(x)$.

---

[3] A real function $f(x)$ is called square-integrable if it satisfies $\int_{x_1}^{x_2} \{f(x)\}^2 dx < \infty$.
[4] For complex functions, we also conjugate one of them, as in Eq. (4.7). Had we considered vectors with complex components, an analogous conjugation would appear in Eq. (4.6).

We thus have a fundamentally important result: just as any vector can be expanded as a linear combination of any complete set of vectors in the vector space, so can any function be expanded as a linear combination of any complete set of functions in the function space. We also know, for both vectors and functions, the forms of the expansion coefficients.[5] Applied to quantum states, these results become some of the most important and pervasive tools in quantum mechanics.

Like vectors and functions, some quantum states are described with a discrete set of expansion coefficients, others with a continuous set. In either case, though, they must obey certain constraints on their mathematical behavior—this restricts them to a particular type of space: *Hilbert space*. (The state spaces in quantum mechanics may be of varying dimensions and contain within them the eigenstates of various operators, but they all must be Hilbert spaces.) Any state in some Hilbert space may be expanded as a linear combination of any complete orthonormal set of states in that Hilbert space.

4.2 Dirac's Bra-ket Notation

4.2.1 Bras and Kets

The great 20th century physicist Paul Dirac introduced into quantum mechanics a notation that is exceptionally useful, and extensively used; it sometimes seems that Dirac's creation has a life and power of its own. We will build this notational system gradually, and point out pitfalls along the way. Please realize from the outset that Dirac notation is *just* notation. Nevertheless, it often appears mysterious and confusing at first. Once you see its simplicity, though, Dirac notation becomes an elegant and powerful tool.

The starting point for Dirac notation is simple: a state, say Ψ, is written as a *ket*, $|\Psi\rangle$. Corresponding to a ket such as $|\Psi\rangle$ is a *bra*, denoted $\langle\Psi|$. We will have much more to say about the relation between bras and kets later. For now, we simply point out that if a bra on the left meets up with a ket on the right, for example, $\langle\Psi|\Psi\rangle$, the result is a bra-ket, or "bracket". Since such meetings are so common, Dirac notation is often called *bra-ket* notation.

As a simple but important first example of bra-ket notation, consider the expansion of a state Ω in a set of states $\{\phi_k\}$:

$$\Omega = \sum_k \alpha_k \phi_k, \tag{4.10}$$

---

[5] For more rigorous treatments see, for example: Jackson (1999), Chapter 2; Arfken (1970), Chapter 9; Kovach (1984), Sections 2.3–2.5.

In bra-ket notation this becomes

$$|\Omega\rangle = \sum_k \alpha_k |\phi_k\rangle. \tag{4.11}$$

The defining relation between a ket state and the corresponding bra state, such as between $|\Omega\rangle$ and $\langle\Omega|$, is

$$|\Omega\rangle = \sum_j \alpha_j |\phi_j\rangle \longleftrightarrow \langle\Omega| = \sum_j \alpha_j^* \langle\phi_j|. \tag{4.12}$$

Equation (4.12) indicates that *if* we write out $\langle\Omega|$ and $|\Omega\rangle$ as superpositions, the correspondence between them is maintained by complex conjugating the expansion coefficients of $\langle\Omega|$ and $|\Omega\rangle$ relative to each other.

Note that Eq. (4.12) defines a correspondence—*not* an equality. It's tempting to think that $\langle\Omega|$ should be expressible *in terms of* $|\Omega\rangle$—that an equation should exist relating $|\Omega\rangle$ and $\langle\Omega|$. But that would be to misconstrue the role of the bra state, which is (as we shall soon see) to facilitate a mathematical operation: taking the scalar product.

4.2.2   Labeling States

There is no new notational form for operators or numbers (such as eigenvalues) in bra-ket notation. Thus, for example, an eigenvalue equation such as $\hat{A}\Psi_k = a_k \Psi_k$ becomes

$$\hat{A}|\Psi_k\rangle = a_k |\Psi_k\rangle. \tag{4.13}$$

We may, however, label bras and kets by any convenient means, so we could have chosen different state labels. A bra or ket is always a quantum state, so it seems a bit redundant to write Ψ in the kets. Instead, we could use the eigenvalues, or just the index k, to uniquely label our states:

$$\hat{A}|a_k\rangle = a_k |a_k\rangle, \text{ or} \tag{4.14}$$

$$\hat{A}|k\rangle = a_k |k\rangle. \tag{4.15}$$

Equations (4.13), (4.14), and (4.15) are all equivalent. But they may not all be equally transparent, at least until you become comfortable with bra-ket notation. In particular, an eigenvalue equation such as Eq. (4.14) could easily be confusing. On the left side of Eq. (4.14) a_k appears inside of a ket, a state. Then the same ket appears on the right side, mysteriously multiplied by a factor a_k, now *not* written inside of a ket. Just what *is* this a_k, anyway?

Such an equation is easily understood *if* one recognizes how the various objects are to be interpreted. In bra-ket notation, if something is written as a bra or ket, such as $|a_k\rangle$ or $\langle a_k|$, it *always* represents a quantum state. With but a few (usually obvious) exceptions, if some object in Dirac notation is not written as a bra or a ket, it is not a state, but some mathematical object such as an operator or eigenvalue.[6] In Eq. (4.14), $|a_k\rangle$ is the k'th eigen*state* of the operator $\hat{A}$, while a_k is the eigen*value* corresponding to the state $|a_k\rangle$.

Because the mathematical objects in Dirac notation may be unfamiliar, the following simple rules may help you to make sense of things.

- A bra-ket is a scalar product—therefore, it is (as we will see), a *number*.
- An operator *does not* act on numbers.
- An operator *does* act on states, that is, bras or kets—the result is a *new state*.

Dirac notation is not well-suited to dealing with specific functional forms of wavefunctions. For example, if $\Psi_j = \kappa_0 \exp(i\alpha_j x)$, with κ a constant, and $\hat{A} = d^2/dx^2$, then

$$\hat{A}\Psi_j = \hat{A}\{\kappa_0 \exp(i\alpha_j x)\} = \frac{d^2}{dx^2}\kappa_0 \exp(i\alpha_j x) = -\kappa_0 \alpha_j^2 \exp(i\alpha_j x). \quad (4.16)$$

But because the label inside of a ket is *just* a label, we certainly would *not* write:

$$\hat{A}|\Psi_j\rangle = \hat{A}|\kappa_0 \exp(i\alpha_j x)\rangle = \frac{d^2}{dx^2}|\kappa_0 \exp(i\alpha_j x)\rangle = -\kappa_0 \alpha_j^2 |\exp(i\alpha_j x)\rangle. \quad (4.17)$$

Often, though, our calculations do not involve specific functional forms, relying instead on the fundamental structure of quantum states—for example, the fact that they are normalized, and that basis states are orthonormal. In such cases, bra-ket notation can be a powerful tool.

4.3 The Scalar Product

4.3.1 Quantum Scalar Products

The scalar product—also called the inner product—of quantum states is analogous to the scalar product of ordinary vectors.[7] The projection of a vector in some direction clearly is of great utility in classical physics. We may, for example, need to find the component of a force or field along some Cartesian axis. What work might we ask a scalar product to perform in quantum mechanics?

---

[6] Important exceptions include state operators (which we will not use), and expressions such as $\langle x|\Psi\rangle = \Psi(x)$. Also, in matrix mechanics, where states are normally represented as row or column vectors (cf. Chapter 6), Dirac notation may also be used.

[7] For both vectors and quantum states, the terms scalar product, inner product, and dot product are synonymous, although the last term is not ordinarily used in quantum mechanics.

The postulates define the fundamental operations we wish to carry out in solving quantum-mechanical problems. Postulate 1 instructs us to find solutions of the time-dependent Schrödinger equation. Postulate 2 tells us that we must obtain the eigenvalues and eigenstates for the operator of interest. Then, by Postulate 3, we expand the system state in a superposition of these eigenstates. The measurement probabilities are then given by the complex squares of the expansion coefficients.

In a superposition, such as $|\Omega\rangle = \sum_k \alpha_k |\phi_k\rangle$, the role of the expansion coefficients, the α_js, is clear: they are "how much" of each $|\phi_j\rangle$ there is in $|\Omega\rangle$. Bra-ket notation does not tell us how to *find* the α_js, but it does provide a means to write them in terms of a set of basis states without committing to the specific form of those states.

To see how this happens, let's look at the simplest possible case. Assume that there are only two basis states, $|\phi_1\rangle$ and $|\phi_2\rangle$. Now, when we take the scalar product of two ordinary vectors, as in Eq. (4.1), both vectors must be expressed in the same coordinate system. Similarly, when we take an inner product in quantum mechanics we work in a common basis for both states. Suppose, then, that we "expand" the basis states $|\phi_1\rangle$ and $|\phi_2\rangle$ in their own ϕ basis. Clearly the expansion coefficients will simply be $\alpha_1 = 1$, $\alpha_2 = 0$ for $|\phi_1\rangle$ and $\alpha_1 = 0$, $\alpha_2 = 1$ for $|\phi_2\rangle$ (analogous to writing, say, the unit vector $\hat{x}$, or $\hat{y}$, in the $\hat{x}$, $\hat{y}$, $\hat{z}$ basis).

In Dirac notation, the scalar product of $|\Omega\rangle$ and $|\phi_1\rangle$ is the bra-ket $\langle\phi_1|\Omega\rangle$.[8] When a bra-ket such as $\langle\phi_1|\Omega\rangle$ appears, summation is implicit, so

$$\langle\phi_1|\Omega\rangle = \Big(1\langle\phi_1| + 0\langle\phi_2|\Big)\Big(\alpha_1|\phi_1\rangle + \alpha_2|\phi_2\rangle\Big)$$
$$= 1\,\alpha_1\langle\phi_1|\phi_1\rangle + 1\,\alpha_2\langle\phi_1|\phi_2\rangle + 0\,\alpha_1\langle\phi_2|\phi_1\rangle + 0\,\alpha_2\langle\phi_2|\phi_2\rangle = \alpha_1.$$
$$(4.18)$$

Here I have used the orthonormality of the $|\phi_j\rangle$s, that is: $\langle\phi_1|\phi_1\rangle = \langle\phi_2|\phi_2\rangle = 1$, and $\langle\phi_1|\phi_2\rangle = \langle\phi_2|\phi_1\rangle = 0$. The scalar product $\langle\phi_1|\Omega\rangle$ is analogous to the scalar product of an arbitrary unit vector, say $\hat{A}$, with a unit *basis* vector, such as $\hat{x}$. For $\hat{A} \cdot \hat{x}$, the result is just A_x, the projection of $\hat{A}$ onto $\hat{x}$. For $\langle\phi_1|\Omega\rangle$, the result is α_1, the projection of $|\Omega\rangle$ onto $|\phi_1\rangle$. Similarly, we could obtain $\langle\phi_2|\Omega\rangle = \alpha_2$.

Now consider a second superposition, $|\gamma\rangle = g_1|\phi_1\rangle + g_2|\phi_2\rangle$. From Eq. (4.12), the scalar product of our two superposition states, $\langle\gamma|\Omega\rangle$, evidently is

$$\langle\gamma|\Omega\rangle = \Big(g_1^*\langle\phi_1| + g_2^*\langle\phi_2|\Big)\Big(\alpha_1|\phi_1\rangle + \alpha_2|\phi_2\rangle\Big) = g_1^*\alpha_1 + g_2^*\alpha_2. \quad (4.19)$$

This scalar product of normalized states is analogous to the scalar product of two unit (but not necessarily basis) vectors.

---

[8] The notation (ϕ_1, Ω), which I will avoid, is also used for the scalar product.

Why should the form of the scalar product that I've employed provide the projection of one state onto another? Specifically, why are the bra coefficients complex conjugated? (That is, why define the bra as in Eq. (4.12)?) Consider the scalar product of a normalized state, such as $|\Omega\rangle$, with itself. Evidently, if this is to constitute the projection of $|\Omega\rangle$ onto itself, we must have $\langle\Omega|\Omega\rangle = 1$. This is certainly satisfied for the form of scalar product we have proposed since, by the normalization condition, $|\alpha_1|^2 + |\alpha_2|^2 = 1$. If, however, we do not complex conjugate when taking the scalar product, the result will not, in general, even be a real number.

For states with continuous eigenvalues, such as two wavefunctions $\psi(x)$ and $\rho(x)$, the scalar product becomes

$$\langle\rho|\psi\rangle = \int_a^b \rho^*(x)\psi(x)dx, \qquad (4.20)$$

where the limits of the integral correspond to the domain of x. We might have expected that the integral in Eq. (4.20) would appear in bra-ket notation as $\int_a^b \langle\rho|\psi\rangle dx$. This, however, is incorrect. In analogy with the discrete eigenvalue case, *when a scalar product appears, such as* $\langle\rho|\psi\rangle$, *integration is implicit in the notation.* Thus, $\langle\rho|\psi\rangle$ corresponds to the *integral* of $\rho^*(x)\psi(x)$.

4.3.2 Discussion

From Section 4.1.2, we know that if some function $f(x)$ is expanded as a linear combination of the orthonormal basis functions $g_j(x)$,

$$f(x) = \sum_k a_k g_k(x), \qquad (4.21)$$

then a_j, the jth expansion coefficient, is given by

$$a_j = \int_a^b g_j^*(x)f(x)dx. \qquad (4.22)$$

But this is just $\langle g_j|f\rangle$, the scalar product of $f(x)$ and $g_j(x)$. Thus, if $|\rho\rangle$ is a basis function, then $\langle\rho|\psi\rangle$ in Eq. (4.20) is really just the expansion coefficient discussed in Section 4.1.2. For the continuous eigenvalue case, then, as for the discrete case, we interpret the scalar product of states in analogy with the scalar product of unit vectors; that is, as the projection of one state onto the other.

So if we expand a state as $|\xi\rangle = \sum_k b_k|\beta_k\rangle$, then $b_j = \langle\beta_j|\xi\rangle$, and this holds regardless of whether $\langle\beta_k|\xi\rangle$ involves a sum, as in Eq. (4.19), or integration, as in Eq. (4.20). Thus, we may write

$$|\xi\rangle = \sum_k b_k|\beta_k\rangle = \sum_k \langle\beta_k|\xi\rangle|\beta_k\rangle. \qquad (4.23)$$

A useful, general, and easily proven relation is

$$\langle \mu | \omega \rangle = \langle \omega | \mu \rangle^*. \tag{4.24}$$

Thus, the probabilities in Eq. (4.23) may be written as $|b_i|^2$, but also as $|\langle \beta_k | \xi \rangle|^2 = \langle \beta_k | \xi \rangle \langle \xi | \beta_k \rangle$.

From Eqs. (4.19) and (4.20), we see that the scalar product of states, a bra-ket, always results in a number, just as does the scalar product of vectors.[9] When you see scalar products in equations, such as in Eq. (4.23), then, keep in mind that they're just numbers, and you can move them around in your calculations like other numbers. Please realize, though, that this applies only to the entire scalar product; for example, $\langle \beta_k | \xi \rangle$ is a number and may be treated as such. But we certainly can't separate this into $\langle \beta_k |$ and $| \xi \rangle$ and treat these objects as numbers; they're *states*.

In Dirac notation, an expectation value (cf. Chapter 3) looks much like a scalar product, but with an operator sandwiched between the states: $\langle A \rangle = \langle \Psi | \hat{A} | \Psi \rangle$. Because expectation values are averages of measurable quantities, they must be numbers. But why, mathematically, should $\langle \Psi | \hat{A} | \Psi \rangle$ be a number?

In Section 4.2.2 I remarked that operators have the effect of changing a quantum state into another quantum state. In light of that fact, we can see that an expectation value is indeed a scalar product. For example, $\hat{A} | \Psi \rangle$ yields a new state: call it $| \psi \rangle$. Then we may write

$$\langle A \rangle = \langle \Psi | \hat{A} | \Psi \rangle = \langle \Psi | \left(\hat{A} | \Psi \rangle \right) = \langle \Psi | \psi \rangle, \tag{4.25}$$

which clearly is just a scalar product—a number. So an expression such as $\langle \Psi | \hat{A} | \Psi \rangle$ *is* a scalar product, but it's *not* the scalar product of the two states that appear explicitly. Rather, it's the scalar product of one of those states *with the new state created by the operator acting on the other of the original states.*

Expectation values occur frequently in quantum mechanics, so it's worth writing one out explicitly, in Dirac notation. Suppose $\hat{G} | g_j \rangle = g_j | g_j \rangle$, and $| \Psi \rangle = \sum_k c_k | g_k \rangle$; then

$$\langle G \rangle = \langle \Psi | \hat{G} | \Psi \rangle = \sum_{m,n} c_m^* \langle g_m | \left(\hat{G} \, c_n | g_n \rangle \right) = \sum_{m,n} c_m^* \langle g_m | \left(g_n c_n | g_n \rangle \right)$$

$$= \sum_{m,n} c_m^* c_n g_n \delta_{m,n} = \sum_n |c_n|^2 g_n. \tag{4.26}$$

---

[9] If, however, the integrand in a scalar product depends on variables other than the variable of integration, the resulting scalar product will also depend on those other variables.

Note that Eqs. (4.26) and (3.6) agree: the expectation value is the sum over all possible measurement results (eigenvalues), each weighted by its probability.

Also in Section 4.2.2, I pointed out that Dirac notation is not well-suited to dealing with specific functional forms of quantum states. Although this may seem restrictive, it often affords great freedom. Because we are not concerned with specific forms, we really don't care whether a scalar product such as $\langle\phi|\Psi\rangle$ involves a summation or an integration, or about the limits of such an integration. Such information is necessary in calculations involving functional forms for states, but not in the sort of abstract calculations done with bras and kets. In Dirac notation, $\langle\phi|\Psi\rangle$ symbolizes the scalar product, regardless of what means would be necessary to actually calculate it.

4.4 Representations

4.4.1 Basics

Without a clear understanding of how eigenstates, superpositions, and representations are related, one cannot truly grasp the structure of quantum states. In Section 2.3.1 we discussed a state $|\Omega\rangle$ that could be expanded in the eigenstates of an operator $\hat{A}$, the $|\Psi\rangle$s, or the (different) eigenstates of $\hat{B}$, the $|\phi\rangle$s:

$$|\Omega\rangle = \sum_{m=1}^{N} c_m |\Psi_m\rangle = \sum_{m=1}^{N} \langle\Psi_m|\Omega\rangle |\Psi_m\rangle, \tag{4.27}$$

$$|\Omega\rangle = \sum_{k=1}^{N} \alpha_k |\phi_k\rangle = \sum_{k=1}^{N} \langle\phi_k|\Omega\rangle |\phi_k\rangle. \tag{4.28}$$

These equations provide two different ways to write $|\Omega\rangle$, that is, two different *representations* of $|\Omega\rangle$. Clearly, the key elements of writing a state in a particular representation are (1) finding the relevant set of basis states (in this case, either the $|\Psi\rangle$s or the $|\phi\rangle$s), and (2) finding the projection of the system state onto each basis state, that is, finding the appropriate scalar products.

Also in Section 2.3.1 we saw that, to determine the probability distribution for measurements of B in the state $|\Omega\rangle$, we expand $|\Omega\rangle$ in the eigenstates of $\hat{B}$, the $|\phi\rangle$s. The motivation for so doing was that the probabilities are then easily obtained, and the action of $\hat{B}$ is simple because it acts on its eigenstates. Of course, this same procedure would serve to determine the probability distribution of B for any state.

Whenever we write a state as a particular superposition, such as in Eqs. (4.27) and (4.28) (or, for that matter, as an eigenstate), we are choosing a representation. Our choice is typically made for reasons similar to those employed above: to enable us to deal with the action of some operator on the state. In Chapter 5 we will see important quantum-mechanical operators that do *not* represent observable quantities. For these, also, we typically choose a representation in which the action of the operator is simple.

4.4.2 Superpositions and Representations

Suppose that, rather than $|\Omega\rangle$, our system is in the state $|\phi_j\rangle$, the j'th eigenstate of $\hat{B}$. The "expansion" of $|\phi_j\rangle$ in the ϕ representation consists, obviously, of only one term: $|\phi_j\rangle$ itself. Of course, we could expand $|\phi_j\rangle$ in the eigenstates of $\hat{A}$: $|\phi_j\rangle = \sum_g \beta_g |\Psi_g\rangle$. A question then arises: is $|\phi_j\rangle$ a superposition, or isn't it?

The answer is: it's both. When expressed in the ϕ representation, $|\phi_j\rangle$ is a single basis state. But in the Ψ representation, $|\phi_j\rangle$ is a superposition. Referring to $|\phi_j\rangle$ as a superposition—or not—has meaning only if it's clear what representation we're dealing with.

This is a fundamentally important and often misunderstood point, so I'll elaborate a bit. If a state is expanded in the eigenstates of some operator, and if that "expansion" consists of only one term, then the state is not a superposition *in that representation*. But if there is more than one term in the expansion, the state *is* a superposition *in that representation*.[10] Although in physics we may refer to some quantum state as "a superposition," please realize that this designation generally makes sense *only* with respect to some specific representation, that is, with respect to expansion over some specific set of basis states.

A simple example makes things more concrete. Consider an infinite potential well: a particle of mass m in a (one-dimensional) region where the potential energy is zero for $0 < x < L$, and infinite elsewhere (see Fig. 4.2). Classically, this corresponds to a particle subject to zero force for $0 < x < L$, and infinite force at $x = 0$ and $x = L$—thus, this system is often called a "particle-in-a-box."

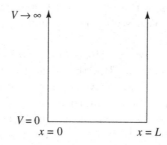

Fig. 4.2 The infinite potential well, or "particle-in-a-box."

---

[10] Often, a state is a superposition in *every* representation, that is, no representation exists in which the state may be written as a single basis state.

In Section 9.2, the eigenstates and eigenvalues of the Hamiltonian operator $\hat{H}$ for the particle-in-a-box are found to be

$$\psi_n(x) \;\; = \;\; \sqrt{2/L} \; \sin\left(n\pi x/L\right), \tag{4.29}$$

$$E_n \;\; = \;\; \frac{n^2\pi^2\hbar^2}{2mL^2}. \tag{4.30}$$

So, the particle-in-a-box problem is solved, right? Yes—and No! Although we have obtained the energy eigenstates[11] (and eigenvalues), these are only a set of basis states. From this set, an *infinite set* of other, legitimate particle-in-a-box states could be formed, by superposition. And, unlike the energy eigenstates, such superpositions will be time dependent (as we will see in Chapter 11).

Fully understanding the structure of the quantum states for even this simple system involves considerable subtlety—but it also leads to considerable insight. Let's begin with an eigenstate of $\hat{H}$, say ψ_2.[12] Expanded in the H representation, we simply have ψ_2 itself. The expansion coefficient is 1, as is the probability: we will, with certainty, obtain $E_2 = 2\pi^2\hbar^2/mL^2$ upon measurement of the energy. Of course, similar comments apply to any one of the ψ_ns.

Recall that each possible position, each x, corresponds to an eigenstate of $\hat{x}$. We see that in Eq. (4.29) the ψ_ns are already written in position representation, that is, as functions of x. Moreover, the ψ_ns are superpositions in the position representation—else they each would consist of only a single x value. Note that the eigenstates and eigenvalues for $\hat{H}$ are discrete for this system, while those for $\hat{x}$ are continuous—an entirely legitimate and very common situation.

Now consider a superposition of the $\psi_n(x)$s, such as

$$\Phi(x) = \frac{1}{\sqrt{2}}\Big\{\psi_2(x) + \psi_4(x)\Big\} = \frac{1}{\sqrt{2}}\Big\{\sin(2\pi x/L) + \sin(4\pi x/L)\Big\}. \tag{4.31}$$

A state's representation is determined by the basis, the eigenstates, over which it's expanded. That's a rather odd statement in this case: though expanded in the H representation, the eigenstate constituents are written as superpositions in the x representation. So, which representation is $\Phi(x)$ written in?

In Eq. (4.31) there is just one coefficient for each eigenstate of $\hat{H}$. There are, however, two coefficients for each eigenstate of $\hat{x}$: at each value of x there is one contribution from $\psi_2(x)$, and another from $\psi_4(x)$. Thus, $\Phi(x)$ is written in the H representation. Such cases are common: to time

---

[11] The states in Eq. (4.29) are solutions to the time *independent* Schrödinger equation. These are sufficient for present purposes, and are what is found in most texts.

[12] For now, I avoid writing wavefunctions, such as the $\psi_j(x)$s, in Dirac notation, as subtle notational issues arise when doing so.

evolve quantum systems we often expand position-space states in the H representation (see Chapter 11).

Clearly, $\Phi(x)$ is an eigenstate of neither $\hat{x}$ nor $\hat{H}$, that is, $\Phi(x)$ is not a single basis state in either the position (x) or energy (H) representations. In fact, there is *no* representation of particle-in-a-box states in which $\Phi(x)$ is a single basis state. If this seems surprising, it shouldn't.

As we'll see in Chapter 5, corresponding to any Hermitian operator (and therefore, by Postulate 2, to any observable) is a complete set of eigenstates—the basis states of that representation. Although we can build any possible *system state* by superposing such eigenstates, such basis states, the concept of a system state is otherwise independent from that of a basis state, or eigenstate. The system state *could* be just a single eigenstate, but it *need* not be. Therefore, the system need not be in an eigenstate of the operator of interest—nor, indeed, of *any* Hermitian operator.

4.4.3   Representational Freedom

A crucial point should, by now, be clear: there is only *one* quantum state of the system,[13] and this quantum state determines the probability distributions for all of the relevant observables. Changing *representations* does not change the *state*. In the example of Section 4.4.2, the ψ_ns are eigenstates of $\hat{H}$ but not of $\hat{x}$, and that fact is not altered merely by changing representations. Restated, each ψ_j remains an eigenstate of $\hat{H}$, even if we choose to write it as a superposition of the eigenstates of $\hat{x}$.

We may choose to work with the system state, which embodies information on the system's various observables, in various representations. As an abstract entity, however, the state itself is representation-free—a codification of information about the quantum system that exists independently of any language, any representation, we choose to write it in. In Eqs. (4.27) and (4.28), for example, the state $|\Omega\rangle$ is written in two different representations. But we could, of course, simply use the ket $|\Omega\rangle$, a representation-free object, to symbolize the state. The situation is similar for scalar products: although they are valid in various representations, the scalar product itself is representation-free.

As a crude but illustrative analogy, imagine discussing a dog with either an American or a Frenchman. In one case we use the word *dog*; in the other, the word *chien*. Either word is perfectly legitimate, although one or the other is more convenient, depending on the audience. However, the actual animal, a physical entity, obviously exists quite apart from the word we use to represent it; in fact, it exists even if we choose not to represent it in words at all.

Dirac notation capitalizes on the representational freedom of quantum mechanics; one way to see this is through expectation values. In Eq. (4.26),

---

[13] There are infinite *possible* system states, but only one in any particular instance.

I calculated the expectation value $\langle G \rangle = \langle \Psi | G | \Psi \rangle$ by expanding the system state in the eigenstates of $\hat{G}$, the operator of interest. This is the usual way to calculate an expectation value, but the expression $\langle \Psi | \hat{G} | \Psi \rangle$ doesn't instruct us to do this. Like states, scalar products are not tied to any representation, and $\langle \Psi | \hat{G} | \Psi \rangle$ simply instructs us, without reference to a particular representation, to take the scalar product of $\langle \Psi |$ and $\hat{G} | \Psi \rangle$.

How could one calculate $\langle \Psi | \hat{G} | \Psi \rangle$ in a representation (basis) other than that formed by the eigenstates of $\hat{G}$? We could write out $| \Psi \rangle$ in this other, "new" basis, and then transform this expansion into the G representation, essentially reproducing Eq. (4.26). Alternatively, by determining the action of $\hat{G}$ on each of the new basis states, we could work directly in the new representation (we will revisit operators in alternate representations in Chapters 5 and 6). As an abstract entity, however, $\langle \Psi | \hat{G} | \Psi \rangle$ exists and is unchanged regardless of what representation we choose to work in—regardless, in fact, of whether we choose any representation at all. The bra-ket expression $\langle \Psi | \hat{G} | \Psi \rangle$ symbolizes $\langle G \rangle$ without commitment to any representation.

Students typically first see expectation values in the context of wavefunctions. Given a wavefunction $\Psi(x)$, and an operator $\hat{A}$, the expectation value $\langle A \rangle$ is

$$\langle \Psi | \hat{A} | \Psi \rangle = \int_{-\infty}^{\infty} \Psi(x) \hat{A} \Psi(x) dx. \tag{4.32}$$

It's tempting to regard Eq. (4.32) as the *definition* of $\langle A \rangle$ for wavefunctions. But that's not quite right; $\langle A \rangle$ is the scalar product of $\langle \Psi |$ with $\hat{A} | \Psi \rangle$, and (as for $\langle \Psi | \hat{G} | \Psi \rangle$ above) that says nothing about representation. Although the integral in Eq. (4.32) is the conventional expression for $\langle A \rangle$, it implicitly selects the x (position) representation. The bra-ket object $\langle \Psi | \hat{A} | \Psi \rangle$, however, provides a representation-free expression of $\langle A \rangle$.[14]

In Section 3.1.2, we considered $\langle x \rangle$, that is, the case $\hat{A} \to \hat{x}$. There we found that, because $\hat{x} = x$ for wavefunctions (in the x representation), $\langle x \rangle$ has a simple interpretation in terms of probabilities. This simplicity is lost if $\hat{A}$ cannot be treated simply as a variable (like $\hat{x}$). Nevertheless, Eq. (4.32) still holds: the form of $\hat{A}$ will vary with the representation in such a way that the expectation value is preserved.

Note that our discussion has been in keeping with the vector analogy: the scalar product of two vectors is independent of our choice of coordinate system. This, however, does not imply that all coordinate systems are equally easy to work in, and it is often well worth choosing wisely. Similarly, the scalar product of two states is independent of our choice of representation. But, also as for vectors, our choice of representation can affect the ease of calculation.

---

[14] Of course, to actually *calculate* $\langle A \rangle$ we must choose a particular representation to work in, and the x representation is one legitimate choice.

Finally, as abstract mathematical objects, both vectors and their scalar products exist regardless of our choice of coordinate system—indeed, they exist even if we choose no coordinate system at all. The same may be said for quantum states and their scalar products: they exist regardless of our choice of representation—indeed, they exist even if we choose no representation at all.

4.5 Problems

1. Suppose that $|A_1\rangle$, $|A_2\rangle$, and $|A_3\rangle$, on the one hand, and $|B_1\rangle$, $|B_2\rangle$, and $|B_3\rangle$, on the other, form two different complete sets of (orthonormal) states of some quantum system. Suppose, also, that $\hat{F}$ is a Hermitian operator corresponding to some observable quantity F, and that $\hat{F}$ satisfies

$$\hat{F}|A_k\rangle = f_k|A_k\rangle.$$

(i) What possible values could result if a measurement of F is made on an arbitrary state of the quantum system?
(ii) Suppose that the system state is $|\Psi\rangle = \frac{\alpha}{\sqrt{3}}|A_1\rangle + \frac{\beta}{\sqrt{3}}|A_2\rangle + \frac{\gamma}{\sqrt{3}}|A_3\rangle$.
(a) What condition must $\alpha^2 + \beta^2 + \gamma^2$ obey?
(b) What is the probability distribution for measurements of F in the state $|\Psi\rangle$? (Provide the possible measured values and the corresponding probabilities.)
(iii) Now suppose that the system state is $|\Phi\rangle = |B_2\rangle$. What is the probability distribution for measurements of F in the state $|\Phi\rangle$? (You can't get a number; just carry it out as far as you can.)

2. In this problem, $\{|\phi_i\rangle\}$ and $\{|\beta_j\rangle\}$ are sets of orthonormal basis states, while c_1, c_3, and ξ are constants (numbers), and

$$|\psi_j\rangle = \sum_{k=1}^{N} b_{jk}|\phi_l\rangle, \qquad \hat{A}|\psi_n\rangle = a_n|\psi_n\rangle.$$

In the following, Latin indices take on all possible values, while a numeric index, such as in $|\psi_3\rangle$, takes on that particular value.
(i) What sort of object (scalar, state, etc.) is:
$\hat{G}\langle\psi_3|\phi_3\rangle\langle\phi_4|\hat{A}\hat{B}|\psi_5\rangle|a_3\rangle$?
(ii) Simplify Γ *as much as possible*:

$$\Gamma = \langle\beta_0| \, c_1 \, \langle\phi_2| \, \xi\hat{A} \, |\psi_5\rangle\langle\phi_j|\phi_k\rangle\langle\psi_l|\psi_m\rangle \, |\beta_0\rangle \, |G_3\rangle.$$

(iii) What sort of object is Γ?
(iv) Simplify Ξ *as much as possible*:

$$\Xi = \langle\beta_0| \, c_3 \, \langle\phi_2| \, \xi\hat{A} \, |\psi_5\rangle\langle\phi_j|\phi_k\rangle\langle\psi_m|\psi_l\rangle \, |\beta_1\rangle \, |G_2\rangle.$$

5
Operators

In Newton's theory reality was determined by a material point in space and time; in Maxwell's theory, by the field in space and time. In quantum mechanics it is not so easily seen.
The concepts, which [Newton] created, are even today still guiding our thinking in physics, although we now know that they will have to be replaced by others farther removed from the sphere of immediate experience, if we aim at a profounder understanding of relationships.
Albert Einstein[1]

There is much yet to be done in quantum mechanics. Calculating time-independent states and attendant energy eigenvalues is the problem that introductory students are typically most familiar with. But quantum mechanics is far richer than just these problems. If the state of our quantum system evolves in time, we will likely want to calculate that time evolution. We may want to predict measurement results for observables other than energy. We have already seen that the mathematical objects corresponding to the physical process of measuring an observable are operators. Other operators correspond to non-measurement processes, such as time evolution.

Operators play a central role in quantum mechanics. The goal in this chapter is not to treat the wide spectrum of quantum-mechanical operators in its entirety, but rather to discuss the most important classes of operators, and to present a broad overview of why and how operators are used.

We might first ask why operators assume a central role in quantum mechanics, but not in classical mechanics. Consider again the fundamental entities in the two theories. In classical mechanics we specify the state of a particle by its position and momentum. From the knowledge of these physical quantities as a function of time we may calculate, by simply applying the relevant definitions, the values of other dynamical quantities.

In quantum mechanics the fundamental entity is the quantum state, a probability amplitude distribution, which in general does not uniquely determine any dynamical quantities. The connection to physically observable predictions is through the measurement algorithms as developed in the postulates. To investigate the behavior of a quantum-mechanical system,

---

[1] A. Einstein, in Schilpp (1949), pp. 81, 83, 31, 33.

then, *we must investigate the behavior of the quantum state.* Let me try to clarify this statement.

Suppose we are interested in how the behavior of some physical system, for example, a charged body in an electric field, might be changed if it were rotated. In classical physics, we calculate the quantities of interest, such as the object's trajectory, directly (though perhaps not easily!). We can also try to envision how the field might interact with the object after rotation.

In quantum mechanics, we cannot calculate the properties of the object directly. We *must* consider the quantum state as the fundamental entity in our calculations, and determine the effects of the rotation on the state, rather than the object itself. Only then can we extract (probabilistic) physical predictions. While we may try to press our classical intuition into service to imagine what will occur, that intuition is often unreliable in quantum mechanics. We should thus *not* think in terms of what happens to "the object", but rather *we should think in terms of, and calculate, what happens to the quantum state.*

In other words, in quantum mechanics, physical processes—such as measurement, time evolution, or rotation—are "done to" the quantum state. And the way that something is done to the quantum state is with mathematical operations, that is, with operators.

Nevertheless, thinking in terms of particles—physical objects—undergoing processes can be useful and intuitively palatable, and physicists very often do just that. But there are times when we can get into trouble by employing such pictures. You should always realize that the fundamental entity in quantum mechanics is *not* a continuously existing particle (or ensemble of particles), but the quantum state.

Please note that none of the foregoing means that real, continuously existing physical particles cannot exist. It simply means that such particles, if they exist, play no essential role in quantum mechanics.

5.1 Introductory Comments

As is usual in quantum mechanics, the operators that we will deal with are all *linear operators*. If $\hat{L}$ is a linear operator, $|\Psi\rangle$ and $|\Gamma\rangle$ are arbitrary states, and a and b are (possibly complex) constants, then linearity means $\hat{L}$ must satisfy:

$$\hat{L}\left(a|\Psi\rangle + b|\Gamma\rangle\right) = a\hat{L}|\Psi\rangle + b\hat{L}|\Gamma\rangle. \tag{5.1}$$

In words, linearity means that an operator has no effect on constants (numbers), and that its action on states obeys the distributive property. Simple though it is, linearity is of great importance: it determines the basic rules by which an operator may be manipulated. With practice, applying these rules becomes second nature.

Before considering specific classes of quantum-mechanical operators, let's try to get a broad overview of the subject. First, consider a fact that is fundamentally important, but often not obvious: *quantum-mechanical operators must "act" on something*—in particular, they act on either states or other operators.

For example, the commutator of two operators $\hat{A}$ and $\hat{B}$ is defined as,[2]

$$[\hat{A}, \hat{B}] \equiv \hat{A}\hat{B} - \hat{B}\hat{A}. \tag{5.2}$$

Commutators appear frequently in quantum mechanics. To actually *calculate* them, however, it is often essential to realize that the operators involved don't stand on their own; they must always act on a state. For example, suppose the operator $\hat{A}$ is simply the variable x, and $\hat{B}$ is the differential operator d/dx. The interpretation of $[\hat{A}, \hat{B}]$ may be unclear, but now assume that the operators act on a state (wavefunction) $f(x)$. Writing out $[\hat{A}, \hat{B}]f(x)$ explicitly, we have,

$$(\hat{A}\hat{B} - \hat{B}\hat{A})f = \left(x\frac{d}{dx} - \frac{d}{dx}x\right)f = x\frac{df}{dx} - \left(f + x\frac{df}{dx}\right) = -f \neq 0. \tag{5.3}$$

Clearly, the failure to realize that operators act on states could, in such a case, lead to confusion.

A useful way to think about the action of operators on states is that *operators have the effect of changing a quantum state into another quantum state*. An obvious exception to this is the identity operator, $\hat{I}$, which does what you might suspect. For any state $|\Psi\rangle$ the action of $\hat{I}$ is given by:

$$\hat{I}|\Psi\rangle = |\Psi\rangle. \tag{5.4}$$

As a simple example of an operator changing a state, consider the action of an operator $\hat{A}$ on its eigenstates. If the eigenvalue equation for $\hat{A}$ is

$$\hat{A}|\Psi_j\rangle = a_j|\Psi_j\rangle, \tag{5.5}$$

then the effect of $\hat{A}$ on the state $|\Psi_j\rangle$ is simply to multiply $|\Psi_j\rangle$ by the eigenvalue a_j (a number). We may legitimately think of $a_j|\Psi_j\rangle$ as a new state, and thus give it a new name. For example, we could say that $a_j|\Psi_j\rangle \equiv |\Omega_j\rangle$. Nevertheless, $|\Psi_j\rangle$ and $|\Omega_j\rangle$ are so similar that we usually don't introduce new state labels in such a case. Of course, if $\hat{A}$ were to act on a *non*-eigenstate, the new state would not be so simply related to the original state.

---

[2] We will study commutators in Chapter 7.

5.2 Hermitian Operators

5.2.1 Adjoint Operators

In an equation such as $\hat{p}|\phi_k\rangle = p_k|\phi_k\rangle$, the operator $\hat{p}$ clearly acts to the right, on the state $|\phi_k\rangle$. It seems natural to make a similar assumption for scalar products—that is, that $\langle\Psi|\hat{A}|\phi_k\rangle$ really means $\langle\Psi|(\hat{A}|\phi_k\rangle)$, the curved brackets indicating that $\hat{A}$ acts on the state to its right, on the ket.[3] But the bra, $\langle\Psi|$, is also a state, and so could also be acted on by an operator. You may suspect, then, that we will routinely need to deal with operators acting on bra states, but in practice such occasions are rare. There is, however, an important type of operator—the *adjoint* operator—that is *defined by* its action on bras, and to which we now turn our attention.

An operator acting on a state creates another state. I'll denote by $|\xi\rangle$ the new state created by $\hat{A}$ acting on $|\Psi\rangle$; that is, $\hat{A}|\Psi\rangle = |\xi\rangle$. From Section 4.2.1, the correspondence between bras and kets is defined by

$$|\Omega\rangle = \sum_j c_j|\phi_j\rangle \longleftrightarrow \langle\Omega| = \sum_j c_j^*\langle\phi_j|. \tag{5.6}$$

For an arbitrary scalar product, such as $\langle\Psi|\hat{A}|\phi_k\rangle$, we do *not* expect, in general, that $\langle\Psi|(\hat{A}|\phi_k\rangle) = ((\langle\Psi|\hat{A})|\phi_k\rangle = \langle\hat{A}\Psi|\phi_k\rangle$ will hold. For example, for two wavefunctions $\alpha(x)$ and $\beta(x)$, the equation

$$\int \alpha^*(x)[\hat{A}\beta(x)]dx = \int [\hat{A}^*\alpha^*(x)]\beta(x)dx, \tag{5.7}$$

will in general *not* hold. Note that in Eq. (5.7) the operator acts only on the state within the braces, and that it is conjugated when acting to the left.

Suppose, however, that there exists an operator $\hat{A}^\dagger$ whose action to the left, on an arbitrary bra $\langle\Psi|$, produces a new bra that corresponds to the ket produced by $\hat{A}$ acting to the right, on the ket $|\Psi\rangle$. That is,

$$\hat{A}|\Psi\rangle = |\xi\rangle \longleftrightarrow \langle\Psi|\hat{A}^\dagger = \langle\xi|. \tag{5.8}$$

The operator $\hat{A}^\dagger$ is called the *adjoint* of $\hat{A}$. It may also be defined in terms of scalar products: Operating to the left with $\hat{A}^\dagger$ in an arbitrary scalar product produces the same result as operating to the right with $\hat{A}$.

$$\langle\Psi|(\hat{A}|\phi_k\rangle) = ((\langle\Psi|\hat{A}^\dagger)|\phi_k\rangle. \tag{5.9}$$

Note that Eq. (5.9) holds even though $\hat{A}$ and $\hat{A}^\dagger$ act on different states. Equation (5.8) therefore seems more transparent; nevertheless, Eq. (5.9) is a direct consequence of Eq. (5.8).

---

[3] Suppose $\hat{B} = d/dx$; writing $g(x)\hat{B}f(x)$ clearly means to differentiate $f(x)$, not $g(x)$.

Unfortunately, some potentially confusing notational variations exist regarding operators generally, and adjoint operators in particular; the choice of which to use is largely a matter of taste. Some authors may write: $\langle\alpha|\hat{A}|\beta\rangle = \langle\alpha|\hat{A}\beta\rangle$, or $\langle\alpha|\hat{A}^\dagger|\beta\rangle = \langle\hat{A}\alpha|\beta\rangle$. Others write $\langle\alpha|\hat{A}^\dagger|\beta\rangle = \langle\hat{A}^\dagger\alpha|\beta\rangle$. Finally, as in this book, some authors avoid writing operators inside of bras or kets (an exception occurs just before Eq. (5.7)).[4] While this is less explicit than the alternatives, it is arguably also less confusing and ambiguous.

5.2.2 Hermitian Operators: Definition and Properties

In general, the adjoint of an operator, such as $\hat{A}^\dagger$, is different than the original operator, $\hat{A}$. It *could* happen, however, that $\hat{A}^\dagger = \hat{A}$, that is, $\hat{A}$ could be equal to its adjoint. Such operators are called *self-adjoint* or *Hermitian*. Clearly, then, if an operator $\hat{A}$ is self-adjoint, the following holds:

$$\langle\Psi|\big(\hat{A}|\phi_k\rangle\big) = \big(\langle\Psi|\hat{A}^\dagger\big)|\phi_k\rangle = \big(\langle\Psi|\hat{A}\big)|\phi_k\rangle. \qquad (5.10)$$

By its very definition, then, we see an important property of a Hermitian operator: when a Hermitian operator appears in a scalar product it may act on either the bra or the ket without changing the result of the calculation. Thus if a scalar product involves a Hermitian operator, the curved brackets and dagger notation become superfluous and are usually omitted.

The expectation value of any Hermitian operator $\hat{A}$, in *any* quantum state,[5] evidently satisfies:

$$\langle A\rangle = \langle\Psi|\big(\hat{A}|\Psi\rangle\big) = \big(\langle\Psi|\hat{A}\big)|\Psi\rangle, \qquad (5.11)$$

or, in terms of wavefunctions,

$$\langle A\rangle = \int_a^b \Psi^*(x)\hat{A}\Psi(x)dx = \int_a^b \Psi(x)\hat{A}^*\Psi^*(x)dx. \qquad (5.12)$$

From Eqs. (5.11) and (5.12) we see that the expectation value of A equals its complex conjugate, or, in bra-ket form,

$$\langle\Psi|\hat{A}|\Psi\rangle = \big(\langle\Psi|\hat{A}|\Psi\rangle\big)^*. \qquad (5.13)$$

Now, $\langle A\rangle$ is a number, and if a number is equal to its complex conjugate, it must be real. Thus, $\langle A\rangle$ must be real for *any* state. It is a short step to show that all eigenvalues of $\hat{A}$ are real.

---

[4] An example of the first approach appears in Shankar (1980), pp. 28–29; the second approach in Liboff (2003), p. 105; the third approach in Sakurai (1994), Section 1.2.

[5] Equations (5.11) and (5.12) apparently require less stringent conditions on an operator than its being Hermitian, since there the states that appear are some ket and the bra that correponds to that same ket, while the adjoint is defined between *any* bra and *any* ket. Ballentine (1998), Section 1.3, addresses this apparent discrepancy.

What are the *physical* implications of Hermitian operators? Postulate 2 tells us that (1) physically observable quantities are represented by Hermitian operators, and that (2) the eigenvalues of such operators correspond to the possible values of measurements. The conventional (though not necessarily convincing) argument is, then, as follows.[6] Observables must be represented by Hermitian operators because the eigenvalues of Hermitian operators are real, and physical measurements always yield real numbers—never imaginary or complex numbers.

I now list without proof three key properties of Hermitian operators.[7]

- The eigenvalues of a Hermitian operator are real.
- The eigenstates of a Hermitian operator are orthogonal.
- For finite-dimensional spaces, the eigenstates of a Hermitian operator form a complete set.

The third property implies an important result: because the eigenstates of a Hermitian operator form a complete set, any state may be expanded in a linear combination of the eigenstates of any Hermitian operator in that space. For infinite-dimensional spaces the third property presents subtle mathematical problems. Nevertheless, the result of a rigorous analysis is that the eigenstates of a Hermitian operator form a complete set even in such spaces.

A very useful example is the calculation of an expectation value, such as $\langle \Psi | \hat{p} | \Psi \rangle$. I will proceed very deliberately, so as to illustrate the fundamentally important principles that appear in this simple calculation.

Because P is an observable, $\hat{p}$ must be a Hermitian operator; we take its eigenstates to be the $|\phi_k\rangle$s. Let's say that, in the P representation, $|\Psi\rangle = c_1|\phi_1\rangle + c_2|\phi_2\rangle$. Then $\langle\Psi| = c_1^*\langle\phi_1| + c_2^*\langle\phi_2|$, and:

$$\langle\Psi|\hat{p}|\Psi\rangle = \langle\Psi|(\hat{p}|\Psi\rangle) = \left(c_1^*\langle\phi_1| + c_2^*\langle\phi_2|\right)\left(\hat{p}\{c_1|\phi_1\rangle + c_2|\phi_2\rangle\}\right). \quad (5.14)$$

In the first equality the brackets indicate explicitly that we will operate to the right. In the second equality we have expanded $\langle\Psi|\hat{p}|\Psi\rangle$ in the $|\phi_k\rangle$s—the eigenstates of $\hat{p}$. By so doing, the action of $\hat{p}$ becomes trivial: it simply returns the corresponding eigenvalues.

$$\left(c_1^*\langle\phi_1| + c_2^*\langle\phi_2|\right)\left(\hat{p}\{c_1|\phi_1\rangle + c_2|\phi_2\rangle\}\right)$$
$$= \left(c_1^*\langle\phi_1| + c_2^*\langle\phi_2|\right)\left(p_1 c_1|\phi_1\rangle + p_2 c_2|\phi_2\rangle\right). \quad (5.15)$$

Because the $|\phi_k\rangle$s are eigenfunctions of a Hermitian operator they are orthogonal. Moreover, we almost always normalize quantum states, so we

---

[6] Ballentine (1998), Section 2.2.

[7] Proofs of these properties are readily available in textbooks. The second property must be modified for degenerate eigenvalues, but I will not discuss these.

take the $|\phi_k\rangle$s to be orthonormal. Thus, the scalar product of two such states obeys $\langle\phi_j|\phi_k\rangle = \delta_{jk}$. Completing the calculation, then, we have:

$$p_1|c_1|^2\langle\phi_1|\phi_1\rangle + p_1 c_2^* c_1 \langle\phi_2|\phi_1\rangle + p_2 c_1^* c_2 \langle\phi_1|\phi_2\rangle + p_2|c_2|^2\langle\phi_2|\phi_2\rangle$$
$$= p_1|c_1|^2 + 0 + 0 + p_2|c_2|^2, \tag{5.16}$$

which is the two possible measurement values, each weighted by the probability of obtaining that value—in accord with our expectations from Chapter 3. [8]

Simple though it is, this example embodies fundamentally important concepts and principles, such as change of representation, the action of an operator on its eigenstates, and orthonormality. Be sure you fully understand what was done in this example, and why! Failure to do so will likely haunt you later.

5.2.3 Wavefunctions and Hermitian Operators

The time has finally come to introduce some explicit forms for operators. Perhaps the most basic quantities in classical physics are position and linear momentum—corresponding to these in quantum mechanics are the position and momentum operators. The "derivation" of such operators, to the extent that they *can* be derived, is in general a difficult task.[9] I won't derive these fundamentally important operators here, but I will introduce and discuss them.

Suppose a one-dimensional system is represented by a position-space wavefunction, $\Psi(x,t)$. The basis states are the eigenstates of position, and as a result, the position operator in this case simply returns the corresponding eigenvalues, so that $\hat{x} = x$ (see Section 2.3.2). Because the position eigenstates are not eigenstates of momentum, the momentum operator $\hat{p}_x$ is not so simple, taking the form $\hat{p}_x = -i\hbar\,\partial/\partial x$ or, in three dimensions, $\hat{p} = -i\hbar\nabla$.

What if we had chosen to work in the momentum representation? Our wavefunction would then be a function of p_x and t, say $\Phi(p_x,t)$, and the basis states would be the eigenstates of $\hat{p}_x$.[10] In the position representation, we replaced $\hat{x}$ with its eigenvalues, since it was acting on its eigenstates. In the momentum representation, the momentum operator may be replaced by its eigenvalues: $\hat{p}_x = p_x$, since $\hat{p}_x$ now acts on *its* eigenstates. Correspondingly, it is now the position operator that is no longer simple, taking

---

[8] Since $\hat{p}$ is Hermitian, we would have obtained the same result if instead we had chosen to operate to the left with $\hat{p}$ (try it).

[9] Rigorous derivations are available, such as: Ballentine (1998), Chapter 3 and Section 4.1; Jordan (1975). A nice "plausibility" derivation of the linear momentum operator, which manages to avoid group theory while nonetheless being based on similar physical considerations, is given in Landau (1958), pp. 38–39.

[10] $\Phi(p_x,t)$ is the same *state* as $\Psi(x,t)$, but in a different representation. It is *not*, however, the same *function* as $\Psi(x,t)$. This is discussed in more detail in Chapter 12.

the form: $\hat{x} = i\hbar\, \partial/\partial p_x$ (or in three dimensions, $i\hbar\nabla$, where the derivatives are taken with respect to p_x, p_y, and p_z).

Other operators, such as those corresponding to angular momentum and energy (discussed in Chapters 8 and 9, respectively), may be constructed from the position and momentum operators, along with the corresponding classical quantity. For example, if we describe a system using Cartesian coordinates, and if the potential energy depends only on position, then the classical Hamiltonian of a system is simply its total energy:

$$H = K + V(x) = \frac{1}{2}mv_x^2 + V(x) = \frac{p_x^2}{2m} + V(x), \qquad (5.17)$$

where K and $V(x)$ are the kinetic and potential energies, respectively, m is the mass and, for simplicity, we work in only one dimension. We may construct the Hamiltonian *operator* for the corresponding quantum-mechanical system through the substitutions $p_x \rightarrow \hat{p}_x$ and $x \rightarrow \hat{x}$. In position representation:

$$\hat{H} = \frac{\hat{p}_x^2}{2m} + V(x) = \frac{1}{2m}\left(-i\hbar\frac{\partial}{\partial x}\right)^2 + V(x) = \left(\frac{-\hbar^2}{2m}\right)\frac{\partial^2}{\partial x^2} + V(x). \quad (5.18)$$

Because in position representation $\hat{x} = x$, the substitution $x \rightarrow \hat{x}$ changes nothing. This "quantization" procedure works also in momentum representation, but because the explicit forms of $\hat{x}$ and $\hat{p}_x$ are then different, so too will $\hat{H}$ take on a different form. We may now write the time-dependent Schrödinger equation in its general form, as in Postulate 1, and in explicit form for our one-dimensional system, in position representation:[11]

$$i\hbar\frac{\partial}{\partial t}\Psi(x,t) = \hat{H}\Psi(x,t) = \left(\frac{-\hbar^2}{2m}\right)\frac{\partial^2}{\partial x^2}\Psi(x,t) + V(x)\Psi(x,t). \qquad (5.19)$$

Finally, note that because $\hat{H}$ is a Hermitian operator (corresponding to the total energy), there is an associated eigenvalue equation:

$$\hat{H}\Psi_n(x,t) = \left(\frac{-\hbar^2}{2m}\right)\frac{\partial^2}{\partial x^2}\Psi_n(x,t) + V(x)\Psi_n(x,t) = E_n\Psi_n(x,t), \quad (5.20)$$

where $\Psi_n(x,t)$ is the nth eigenstate of $\hat{H}$, and E_n is the nth energy eigenvalue.[12]

---

[11] In essence, Postulate 1 takes the state to depend *only* on t, so the time derivative *cannot* be a partial. In Eq. (5.19), however, the state depends on t *and* on x, so differentiation with respect to t must be indicated with a partial derivative. This sort of notational subtlety occurs frequently when dealing with wavefunctions, for example, when using $\hat{H}$ and $\hat{p}$.

[12] We will revisit Eq. (5.20), the *time-independent Schrödinger equation*, in Chapter 9.

5.3 Projection and Identity Operators

5.3.1 Projection Operators

We have now seen that superposition is pervasive in quantum mechanics. Let's again write down a generic superposition:

$$|\psi\rangle = \sum_j c_j |\xi_j\rangle. \tag{5.21}$$

Suppose Eq. (5.21) holds (i.e. $|\psi\rangle$ can be expanded in the $|\xi_k\rangle$s) even if $|\psi\rangle$ is allowed to be *any* legitimate quantum state of the system. Then Eq. (5.21) constitutes a statement of *completeness*. That is, because any legitimate state can be expanded over the $|\xi_j\rangle$s, the $|\xi_j\rangle$s must constitute a complete set of basis states.[13] In words, Eq. (5.21) says that to expand some arbitrary $|\psi\rangle$ in terms of the $|\xi_j\rangle$s, simply weight each $|\xi_j\rangle$ properly and add them all up.

If we take the $|\xi_j\rangle$s to be orthonormal, the proper weighting is just $\langle\xi_j|\psi\rangle$, the projection of $|\psi\rangle$ onto each $|\xi_j\rangle$. Let's now write our superposition out in a somewhat different form:

$$|\psi\rangle = \sum_j c_j |\xi_j\rangle = \sum_j \langle\xi_j|\psi\rangle|\xi_j\rangle = \sum_j |\xi_j\rangle\langle\xi_j|\psi\rangle. \tag{5.22}$$

Here I've used the fact that the scalar product, $\langle\xi_j|\psi\rangle$, is just a number (it's just c_j), and may be treated as such.

For a single term, such as the $j = 5$ term, we can write:

$$c_5|\xi_5\rangle = \langle\xi_5|\psi\rangle|\xi_5\rangle = |\xi_5\rangle\langle\xi_5|\psi\rangle = \Big(|\xi_5\rangle\langle\xi_5|\Big)|\psi\rangle. \tag{5.23}$$

Bracketing $|\xi_5\rangle\langle\xi_5|$ doesn't really "do" anything other than suggest we view $|\xi_5\rangle\langle\xi_5|$ as an object in its own right. But what is such an object? Evidently it's *not* just a number. The role of a bra is to meet up with a ket and form a scalar product. And the role of an operator is to act on a state to form another state. Thus, $|\xi_5\rangle\langle\xi_5|$ plays the role of an operator: if it meets up with a ket on the right, say $|\alpha\rangle$, the result is the new state $|\xi_5\rangle\langle\xi_5|\alpha\rangle$ (i.e. it is the *state* $|\xi_5\rangle$ weighted by the *number* $\langle\xi_5|\alpha\rangle$). So, while we may think of $|\xi_5\rangle\langle\xi_5|\psi\rangle$ in Eq. (5.23) as the ket $|\xi_5\rangle$ weighted by the scalar product $\langle\xi_5|\psi\rangle$, we may instead think of it as the *operator* $|\xi_5\rangle\langle\xi_5|$, *acting on* the state $|\psi\rangle$.

Clearly, $|\xi_5\rangle\langle\xi_5|$ acting on an arbitrary state $|\alpha\rangle$ yields the $j = 5$ term in the expansion of $|\alpha\rangle$ over the $|\xi_k\rangle$s; that is, it creates a new state consisting simply of $|\xi_5\rangle$ weighted by the projection of $|\alpha\rangle$ onto $|\xi_5\rangle$. Accordingly, we call $|\xi_5\rangle\langle\xi_5|$ a *projection operator*, and denote it $\hat{P}_5$.

---

[13] This is not to be confused with the proposed completeness of quantum mechanics as a description of the world (see Chapter 3).

5.3.2 The Identity Operator

By definition, the identity operator leaves unchanged any state upon which it acts. It would thus seem trivially simple. But in quantum mechanics the term identity operator often refers to much more than simply doing nothing to a state.

What if we were to act on a state $|\psi\rangle$ with not just one projection operator, but with a new operator consisting of the sum of *all* projection operators for a given set of orthonormal basis states? Then we would simply regain Eq. (5.22):

$$\sum_j \hat{P}_j |\psi\rangle = \sum_j |\xi_j\rangle\langle\xi_j|\psi\rangle = |\psi\rangle. \tag{5.24}$$

Thus, the action of the operator $\sum_j \hat{P}_j = \sum_j |\xi_j\rangle\langle\xi_j|$ is just to return the state upon which it acts. In precisely this sense, $\sum_j \hat{P}_j = \hat{I}$, the identity operator.

In another sense, however, $\sum_j \hat{P}_j$ performs a critically important function: it expands a state (in this case $|\psi\rangle$) in a particular set of basis states (in this case, the $|\xi_j\rangle$s). By analogy, if we transform a vector in ordinary, three-dimensional space, *three-space*, into another coordinate system, the vector itself remains unchanged; but we've nonetheless done something very important. The result of translating *War and Peace* from Russian to English is still *War and Peace*, and in that sense, the translator has "done nothing"—but the translator will feel he's done quite a bit!

The identity operator, in the sense used above, is thus really a means to effect a change of basis, or representation. Because it changes the language in which the state is written, but doesn't alter the state itself, we can insert $\hat{I} = \sum_j \hat{P}_j$ wherever convenient. In fact, this operator may act not only on states but also on other operators (discussion of which we will defer to later chapters).

As a simple example of the power of "doing nothing" with the identity operator, suppose we wish to calculate the scalar product $\langle\beta|\Psi\rangle$, but don't know how to do so directly. We do, however, know how to expand $|\beta\rangle$ in the ξ basis, and $|\Psi\rangle$ in the ϕ basis, and we can find the projections of the ξ basis states onto the ϕ basis states. Then we can immediately write

$$\langle\beta|\Psi\rangle = \sum_{j,k}\langle\beta|\xi_j\rangle\langle\xi_j|\phi_k\rangle\langle\phi_k|\Psi\rangle. \tag{5.25}$$

5.4 Unitary Operators

In addition to Hermitian operators, *unitary operators* constitute a fundamentally important class of quantum-mechanical operators.[14] To develop

---

[14] Closely related to unitary operators are anti-unitary operators, which act as unitary operators that also complex conjugate any constants. The time-reversal operator, which we will not further discuss, is probably the most important anti-unitary operator.

the concept of a unitary operator let's first define the inverse of an operator $\hat{U}$ as another operator, $\hat{U}^{-1}$, such that,

$$\hat{U}\hat{U}^{-1} = \hat{U}^{-1}\hat{U} = \hat{I}, \tag{5.26}$$

where $\hat{I}$ is the identity operator (now in the simpler sense of an operator that truly does nothing to a state).[15] By definition, if the inverse of some operator $\hat{U}$ is equal to its adjoint, that is, if

$$\hat{U}^{-1} = \hat{U}^{\dagger}, \tag{5.27}$$

then $\hat{U}$ is a unitary operator. Thus, for such an operator,

$$\hat{U}^{-1}\hat{U} = \hat{U}^{\dagger}\hat{U} = \hat{I}. \tag{5.28}$$

As with any operator, unitary operators have the effect of changing a quantum state into another quantum state. In what follows, primes denote new states created by a unitary operator $\hat{U}$ acting on the states $|\Psi\rangle$ and $|\Omega\rangle$:

$$\hat{U}|\Psi\rangle \equiv |\Psi'\rangle, \quad \text{and} \quad \hat{U}|\Omega\rangle \equiv |\Omega'\rangle. \tag{5.29}$$

Moreover, by definition of an adjoint operator, $\langle\Psi|\hat{U}^{\dagger} = \langle\Psi'|$, and $\langle\Omega|\hat{U}^{\dagger} = \langle\Omega'|$. Now, the scalar product of one state with another (normalized) state gives the projection of one onto the other. Let's examine the scalar product of the primed states, and compare it to that of the unprimed states:

$$\langle\Omega'|\Psi'\rangle = \left(\langle\Omega|\hat{U}^{\dagger}\right)\left(\hat{U}|\Psi\rangle\right) = \langle\Omega|\hat{U}^{\dagger}\hat{U}|\Psi\rangle = \langle\Omega|\hat{U}^{-1}\hat{U}|\Psi\rangle = \langle\Omega|\Psi\rangle. \tag{5.30}$$

So unitary transformations leave the scalar product unchanged, or invariant.

Now let's go back to our old friend, the geometrical vector in three-space. Consider the scalar product between two such vectors $\vec{A}$ and $\vec{B}$ in this space. What can we "do to" our system of coordinate axes and vectors in this case such that we obtain a result analogous to that above for quantum states? That is, what can we do to our coordinate axes and vectors $\vec{A}$ and $\vec{B}$ such that the scalar (dot) product, $\vec{A} \cdot \vec{B}$, is unchanged? Suppose we identically rotate all vectors in our system (except the basis vectors). Clearly, this cannot alter the vectors' lengths. Moreover, regardless of the new orientation of these vectors with respect to the fixed coordinate system, *their orientation with respect to each other remains the same*, and thus their dot product remains the same.

So rotations in three-space preserve relations between vectors, while unitary transformations have a similar effect, in the sense that they leave scalar products of quantum states unchanged. Thus we may think of a unitary transformation as a rotation in Hilbert space—the abstract, mathematical space in which quantum states live.

---

[15] Some authors use 1 rather than $\hat{I}$. You should realize that a product of operators, by itself, can't be a number!

Unitary transformations are used in part to investigate conservation principles and symmetries. To pursue such investigations in depth would require us to venture into the wild and woolly world of group theory, which we will not do.[16] We can nevertheless examine some of the most important basics regarding unitary operators that may prove useful later.

We have seen that the physical meaning of Hermitian operators is that they correspond to observables. What is the physical meaning of unitary operators? For our purposes, *unitary operators effect transformations of quantum-mechanical systems in space and time.* Such transformations include rotations, boosts (changes in velocity), translations in time, and translations in space (along an axis). Unitary transformations may be carried out on either the states or the observables (Hermitian operators) with equivalent results.[17]

To illustrate this, consider the behavior of the expectation value $\langle \Psi | \hat{A} | \Psi \rangle$ under a unitary transformation. Proceeding as in Eq. (5.30), we write,

$$\langle \Psi' | \hat{A} | \Psi' \rangle = \left(\langle \Psi | \hat{U}^\dagger \right) \hat{A} \left(\hat{U} | \Psi \rangle \right) = \langle \Psi | \left(\hat{U}^\dagger \hat{A} \hat{U} \right) | \Psi \rangle. \qquad (5.31)$$

Equation (5.31) expresses the fact that we may regard the transformed scalar product *either* as the expectation value of A in the transformed state $|\Psi'\rangle$, *or* as the expectation value of the *transformed operator* $\hat{U}^\dagger \hat{A} \hat{U}$ in the *un*transformed state $|\Psi\rangle$. This illustrates the fact that we may transform *either* the states, by operating with $\hat{U}$ (and $\hat{U}^\dagger$), *or* the observables, with a transformation of the form $\hat{U}^\dagger \hat{A} \hat{U}$.

Although for our purposes unitary operators are of secondary importance, one such operator appears routinely in quantum mechanics, and plays a central role in what follows: the *time-evolution operator*, $\exp[-i\hat{H}(t - t_0)/\hbar]$, where $\hat{H}$ is the Hamiltonian operator, t is the time, and t_0 is the initial time.[18] The role of the time evolution operator is to propagate, or "time evolve," quantum states from initial time t_0 to time t, and therein lies its great utility. We will have much more to say about the time evolution operator later.

5.5 Problems

1. Consider a quantum system which requires only two basis states to span the space. For the operator $\hat{G}$, the basis states are $|G_1\rangle$ and $|G_2\rangle$.

---

[16] Unitary operators and transformations are discussed further, however, in Appendix D.

[17] These two equivalent methods of carrying out a unitary transformation, that is, transforming the states or the observables, are often referred to as active and passive transformations, respectively.

[18] This form of the time-evolution operator is derived by separating variables in the time-dependent Schrodinger equation. This requires that $\hat{H}$ be independent of time; thus, this form generally holds only for such $\hat{H}$s (cf. Chapter 11).

A mythical "black box" prepares the system in the state $|G_2\rangle$. Another operator, $\hat{A}$, has eigenstates $|a_1\rangle$ and $|a_2\rangle$ that span the space, with eigenvalues a_1 and a_2, respectively.

The eigenstates of the Hamiltonian operator, $\hat{H}$, are $|E_1\rangle$ and $|E_2\rangle$. These are needed to time evolve the system; I'll explain below. Your goal is to calculate the *time-dependent* probability distribution for measurements of A. I'm going to help you through this.

Use bra-ket notation wherever possible, including for the expansion coefficients.

(i) We will develop time evolution in a later chapter. For now, I'll simply provide you with what's needed. Time evolution is carried out by the action of an operator $\hat{U}(t)$, where: $\hat{U}(t) = e^{-i\hat{H}t/\hbar}$. It can be shown that *if* $\hat{U}(t)$ acts on, say, the jth eigenstate of $\hat{H}$, with eigenvalue E_j, then the action of $\hat{U}(t)$ is simply $\hat{U}(t) \longrightarrow e^{-iE_j t/\hbar}$.

Now expand the initial state $|G_2\rangle$ in a basis that will allow you to time evolve it.

(ii) Call the time-dependent system state $|\psi(t)\rangle$. Write down $|\psi(t)\rangle$ in terms of the expansion of Part (i). Simplify $|\psi(t)\rangle$ as much as possible—remember that the point is to replicate what a practicing physicist would actually do, that is, to obtain something that you could actually *calculate*.

(iii) Write down eigenstate expansions that will allow you to calculate the probability distribution for $\hat{A}$ from $|\psi(t)\rangle$.

(iv) Rewrite $|\psi(t)\rangle$ using the result of Part (iii).

(v) Find the probability distribution for a_1 and a_2.

(vi) Write a clear and coherent summary (in words!) describing what you did in Parts (i) through (v), and why.

2. In Chapter 2 we first introduced the idea that observable quantities are represented by Hermitian operators. But it is worth examining the meaning of this statement.

   In what way does the mathematical process of acting on a state with a Hermitian operator correspond to the physical process of making a measurement? In what way does it *fail* to correspond to making a measurement? (Hint: you will probably need to consider the changes brought about by measurement, per Postulate 3, to answer this question.)

3. In general, the fact that the average value of some quantity is real (not imaginary or complex) certainly doesn't mean that each individual value over which the average is taken must be real.

   If an operator is Hermitian, the expectation value for the operator is real for *any* quantum state. This statement alone is sufficient to guarantee that any individual eigenvalue of the operator is real. Explain why this is so.

   Please refrain from reproducing proofs, which are easily found in texts, that the eigenvalues of Hermitian operators are real. Rather, think carefully about the second sentence of this question. You should be able to answer the question convincingly using little or no mathematics.

4. Suppose $\hat{D}|\delta\rangle = |\delta'\rangle$, where $|\delta\rangle$ is an arbitrary state. Write an equation indicating the action of the operator $\hat{D}^\dagger$ on the bra state corresponding to $|\delta\rangle$.

5. If $\hat{B}$ is a Hermitian operator, then any state may be expanded in the eigenstates of $\hat{B}$. What property of Hermitian operators does this reflect?

6. Consider a unitary operator $\hat{U}(\gamma) = e^{-i\hat{G}\gamma/\hbar}$. Assume that $\hat{G}|\phi_k\rangle = g_k|\phi_k\rangle$, and take $|\Psi\rangle$ to be a superposition of two eigenstates of $\hat{G}$, i.e., $|\Psi\rangle = c_\alpha|\phi_\alpha\rangle + c_\beta|\phi_\beta\rangle$.

(i) Determine whether the inner product $\langle\Psi|\Psi\rangle$ is preserved under the indicated transformations.

(a) $\langle\Psi|\Psi\rangle \longrightarrow \langle\Psi|\hat{U}(\gamma)|\Psi\rangle$
(b) $\langle\Psi|\Psi\rangle \longrightarrow \langle\Psi|\hat{U}^\dagger(\gamma)|\Psi\rangle$
(c) $\langle\Psi|\Psi\rangle \longrightarrow \langle\Psi|\hat{U}^\dagger(\gamma)\hat{U}(\gamma)|\Psi\rangle$

(ii) It is sometimes said that "inner products are preserved (left unchanged) by unitary operators." Given your results in Part (i), clarify/explain what is meant by this statement.

(iii) Perhaps the most important unitary operator is the time evolution operator, denoted $\hat{U}(t)$, which advances a quantum state forward in time. Explain why your answers to Parts (i) and (ii) make sense *physically*, for the case of time evolution. (Why would you expect them to hold, based on what is being time evolved in Part (i)?)

7. The operators $\hat{A}$ and $\hat{B}$, with eigenvalues α_j and β_j, respectively, share eigenstates. The orthonormal set $\{|j\rangle\}$, where $j = 1, 2, 3..., \infty$, comprises these eigenstates. Suppose the system state is $|\gamma\rangle$, and that $|\gamma\rangle$ is *not* an eigenstate of $\hat{A}$ or $\hat{B}$.

(i) Write down three equations that concisely state the information provided in the preceding paragraph.

(ii) Calculate $\langle\gamma|\hat{A}|\gamma\rangle$. Be explicit.

(iii) Calculate $\langle\gamma|\hat{C}|\gamma\rangle$, where $\hat{C} = \hat{A}\hat{B}$. Be explicit. So far, we are taking operators to act only to the right. We regard $\hat{C}$ as meaning that first $\hat{B}$ acts, and then $\hat{A}$.

(iv) The states $|\Delta_1\rangle$ and $|\Delta_2\rangle$ are (different) superpositions of the $|j\rangle$s. Write out equations that state this (and thus define the expansion coefficients). Then calculate explicitly $\langle\Delta_1|\hat{A}|\Delta_2\rangle$ and $\langle\Delta_2|\hat{A}|\Delta_1\rangle$.

(v) Suppose that the set $\{|J\rangle\}$, where J runs from $-\frac{3}{2}$ to $\frac{3}{2}$ in integer steps, forms a complete set of basis states for some system. Also,

$$\hat{G}|J\rangle = g_J|J\rangle, \text{ and} \tag{5.32}$$

$$|\Psi\rangle = \sum_{J=-3/2}^{3/2} c_J|J\rangle, \tag{5.33}$$

where $g_J = \hbar/J$, and the *unnormalized* c_Js are proportional to $\sqrt{|J|}$. Obtain the normalized c_Js, and then calculate $\langle\Psi|\hat{G}|\Psi\rangle$.

6

Matrix Mechanics

Heisenberg criticized Schrödinger's approach... In a letter to Pauli he e⸱
"The more I ponder about the physical part of Schrödinger's theory, the
gusting it appears to me." Schrödinger was not less outspoken about He⸱
theory when he said: "I was discouraged, if not repelled, by what appea⸱
a rather difficult method of transcendental algebra, defying any visualiz⸱
Max Jammer[1]

Built on the labor of many physicists, quantum mechanics w⸱
over many years. A lack of coherence plagued "the old quant⸱
ory" until finally, in 1926 and 1927, quantum mechanics was
a coherent form. Almost simultaneously, two separate formalis⸱
introduced: Erwin Schrödinger's wave mechanics and Werner Hei⸱
matrix mechanics.

Wave mechanics was warmly welcomed—it was based on the
mathematics of differential equations, and seemed amenable t⸱
standable physical pictures. But Heisenberg's matrix mechanics w⸱
abstract, and seemed to have little clear connection to a physic⸱
pretation. The original matrix mechanics was far more opaque t⸱
version which eventually emerged, and that is widely used today⸱
theless, even this distilled version can seem abstract and unclear,
on first acquaintance—a hurdle I hope to lower in this chapter.

6.1   Elementary Matrix Operations

6.1.1   Vectors and Scalar Products

Before developing the matrix formulation of quantum mechanics-
mechanics—we first briefly review some elementary definitions a⸱
ations with vectors and matrices. You may have covered this g⸱
elementary algebra, but it's worth presenting here because it pr⸱
concise and convenient review and reference, and because it's sp⸱
intended to lay the groundwork for our subsequent development o⸱
mechanics.

---

[1] Jammer (1966), p. 272.

Consider a vector $\vec{V}$ in three-space. Writing out $\vec{V}$ in a basis comprised of the unit vectors $\hat{x}$, $\hat{y}$, and $\hat{z}$, we have: $\vec{V} = V_x\hat{x} + V_y\hat{y} + V_z\hat{z}$. But if it's understood from the outset that we are working in the $\hat{x}$, $\hat{y}$, $\hat{z}$ basis, it's unnecessary to restate that fact when writing out $\vec{V}$. In that case, $\vec{V}$ may be fully specified by specification of its components along each basis vector.[2] Thus, I could write $\vec{V}$ as either a 1×3 (one row, three column) *row vector*, or as a 3×1 (three row, one column) *column vector*:

$$\vec{V} = (V_x, V_y, V_z), \quad \text{or} \quad \vec{V} = \begin{pmatrix} V_x \\ V_y \\ V_z \end{pmatrix}. \tag{6.1}$$

Now introduce a second vector, say $\vec{U}$. The scalar (dot) product of $\vec{U}$ and $\vec{V}$ is: $\vec{U} \cdot \vec{V} = U_xV_x + U_yV_y + U_zV_z$. We define multiplication of row and column vectors in consonance with the scalar product, that is,[3]

$$\vec{U} \cdot \vec{V} \equiv (U_x, U_y, U_z) \begin{pmatrix} V_x \\ V_y \\ V_z \end{pmatrix} = \sum_{j=x,y,z} U_jV_j = U_xV_x + U_yV_y + U_zV_z.$$

$$\tag{6.2}$$

The concept of a row vector, a column vector, and the scalar product of two vectors may be immediately generalized to an n-dimensional space, where n is some arbitrary positive integer. Then the row vector $\vec{U}$ is: $\vec{U} = (U_1, U_2, \ldots, U_n)$, and similarly for the corresponding column vector. The scalar product of two n-dimensional vectors $\vec{U}$ and $\vec{V}$ is then: $\vec{U} \cdot \vec{V} = \sum_{j=1}^{n} U_jV_j$.

6.1.2 Matrices and Matrix Multiplication

A *matrix* has multiple rows and columns. In the matrix formulation of quantum mechanics we will be concerned exclusively with *square* matrices—those with an equal number of rows and columns. The elements of a matrix are denoted by a double subscript that indicates the row and column of the element, in that order. Thus, an $n \times n$ matrix $\hat{M}$ is written:

$$\hat{M} = \begin{pmatrix} M_{11} & M_{12} & \cdots & M_{1n} \\ M_{21} & M_{12} & \cdots & M_{2n} \\ \vdots & \vdots & \ddots & \vdots \\ M_{n1} & M_{n2} & \cdots & M_{nn} \end{pmatrix}. \tag{6.3}$$

---

[2] This point was also made in Chapter 4.

[3] Note that it is *not* legitimate to write $\vec{U} \cdot \vec{V}$ as: $\begin{pmatrix} U_x \\ U_y \\ U_z \end{pmatrix} (V_x, V_y, V_z)$. Such an expression yields, by definition, an *outer product*, which is a *matrix* (in this case, of dimension 3×3).

Multiplication of the $n \times n$ matrix $\hat{M}$ by an $n \times 1$ column vector $\vec{V}$ on *the right* results in another n-dimensonal column vector, which I'll denote $\vec{V}'$. In particular, $\sum_{j=1}^{n} M_{ij} V_j = V_i'$. Illustrating with the simplest case of $n = 2$:

$$\begin{pmatrix} M_{11} & M_{12} \\ M_{21} & M_{22} \end{pmatrix} \begin{pmatrix} V_1 \\ V_2 \end{pmatrix} = \begin{pmatrix} M_{11}V_1 + M_{12}V_2 \\ M_{21}V_1 + M_{22}V_2 \end{pmatrix} = \begin{pmatrix} V_1' \\ V_2' \end{pmatrix}. \tag{6.4}$$

Similarly, multiplication of the $n \times n$ matrix $\hat{M}$ by a $1 \times n$ row vector $\vec{V}$ *on the left* results in another n-dimensonal row vector $\vec{V}'$. Explicitly, $\sum_{i=1}^{n} V_i M_{ij} = V_j'$, so for the $n = 2$ case,

$$(V_1, \ V_2) \begin{pmatrix} M_{11} & M_{12} \\ M_{21} & M_{22} \end{pmatrix} = (V_1 M_{11} + V_2 M_{21}, \ V_1 M_{12} + V_2 M_{22})$$

$$= (V_1', \ V_2'). \tag{6.5}$$

Finally, suppose we wish to multiply $\hat{M}$ on the left with another $n \times n$ matrix $\hat{F}$, on the right. The result is a new $n \times n$ matrix, which we'll denote $\hat{G}$. An element of $\hat{G}$ is given by: $\sum_{k=1}^{n} M_{ik} F_{kj} = G_{ij}$. For the $n = 2$ case,

$$\begin{pmatrix} M_{11} & M_{12} \\ M_{21} & M_{22} \end{pmatrix} \begin{pmatrix} F_{11} & F_{12} \\ F_{21} & F_{22} \end{pmatrix} =$$

$$\begin{pmatrix} M_{11}F_{11} + M_{12}F_{21} & M_{11}F_{12} + M_{12}F_{22} \\ M_{21}F_{11} + M_{22}F_{21} & M_{21}F_{12} + M_{22}F_{22} \end{pmatrix} = \begin{pmatrix} G_{11} & G_{12} \\ G_{21} & G_{22} \end{pmatrix} \tag{6.6}$$

6.1.3 Vector Transformations

Now suppose we wish to "do something to" the n-dimensional vector $\vec{V}$. Perhaps the simplest example is to again set $n = 2$ and rotate $\vec{V}$ through an angle θ, as in Fig. 6.1. Let's call the new, rotated vector $\vec{V}'$. Careful analysis of Fig. 6.1 shows that V_x' depends on *both* components of the original vector, V_x and V_y (and, of course, on θ), and similarly for V_y'; specifically:

$$V_x' = V_x \cos\theta - V_y \sin\theta, \tag{6.7}$$
$$V_y' = V_x \sin\theta + V_y \cos\theta. \tag{6.8}$$

This is a general feature of vector transformations: any *one* component of the transformed vector will in general depend on *all* components of the original vector.

A natural means to effect such a transformation is with a matrix. To carry out the rotation of Fig. 6.1, for example, we may define a rotation

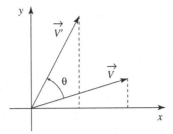

Fig. 6.1 Rotation of a vector $\vec{V}$ through an angle θ; the dashed lines indicate the projections onto the x axis.

matrix $\hat{R}(\theta)$ such that

$$\hat{R}(\theta)\vec{V} = \begin{pmatrix} \cos\theta & -\sin\theta \\ \sin\theta & \cos\theta \end{pmatrix} \begin{pmatrix} V_x \\ V_y \end{pmatrix} = \begin{pmatrix} V'_x \\ V'_y \end{pmatrix}. \tag{6.9}$$

Evidently matrices are well suited for carrying out such transformations.

Before returning to quantum mechanics, I emphasize that most vectors and matrices in this section make no explicit reference to a basis. Rather, they are *implicitly* written with respect to the $\hat{x}$, $\hat{y}$, $\hat{z}$ basis.

6.2 States as Vectors

As for an ordinary vector, a quantum state written in a basis with discrete basis states is conveniently represented as rows, or columns, of coefficients. For an ordinary vector, these coefficients are the vector's components, its projections onto the chosen basis vectors. For a quantum state, the coefficients are the expansion coefficients, the projections of the state onto the chosen basis states. For both cases, it must be understood what basis we are working in.

Unlike the components of ordinary vectors, the expansion coefficients for a quantum state are, in general, complex. The correspondence between bras and kets is: $|\psi\rangle = \sum_j c_j |\phi_j\rangle \longleftrightarrow \langle\psi| = \sum_j c_j^* \langle\phi_j|$. In matrix mechanics a ket is represented as a column whose elements are the expansion coefficients, the c_ks, while the corresponding bra appears as a row whose elements are the c_k^*s. Thus, the scalar product of a two-component state $|\psi\rangle$ with itself is

$$\langle\psi|\psi\rangle = (c_1^*, \; c_2^*) \begin{pmatrix} c_1 \\ c_2 \end{pmatrix} = |c_1|^2 + |c_2|^2. \tag{6.10}$$

A few simple but important observations are in order.

First, note that if $\vec{U}$ and $\vec{V}$ are two ordinary vectors, then $\vec{U} \cdot \vec{V} = \vec{V} \cdot \vec{U}$ (cf. Eq. (6.2)). But this clearly does *not* hold for two quantum states, say $|\psi\rangle$

and $|\xi\rangle$. That is, $\langle\psi|\xi\rangle = \langle\xi|\psi\rangle^* \neq \langle\xi|\psi\rangle$. This shouldn't be too surprising: for the continuous eigenvalue case, the scalar product of two functions $f(x)$ and $g(x)$ is

$$\langle f|g\rangle = \int_{x_0}^{x_1} f^*(x)g(x)dx \neq \int_{x_0}^{x_1} g^*(x)f(x)dx. \tag{6.11}$$

The order of the bra and ket in a scalar product matters.

Note also that when states are represented as vectors, the relation between a basis state and a superposition is particularly clear. Because the vector's components are just the expansion coefficients (or their complex conjugates), then if the state is a basis state in the chosen basis, all components must be zero except for one of them, which will be 1. If the state is a superposition in the chosen basis, then at least two entries must be non-zero. Suppose the $\{\phi_j\}$s constitute an n-dimensional basis; then $\langle\phi_j|\phi_k\rangle = \delta_{jk}$ expresses orthonormality. It's easy to see how this statement is manifested in matrix mechanics:

$$\langle\phi_j|\phi_k\rangle = \left(0,\ldots,c_j^* = 1,\ldots,0\right) \begin{pmatrix} 0 \\ \vdots \\ c_k = 1 \\ \vdots \\ 0 \end{pmatrix} = \left\{ \begin{array}{l} 0 \ \text{if} \ j \neq k \\ 1 \ \text{if} \ j = k \end{array} \right\}, \tag{6.12}$$

where j and k can take on any value from 1 to n.

6.3 Operators as Matrices

6.3.1 An Operator in Its Eigenbasis

In Section 6.1.3 I illustrated how matrices transform vectors. We've also seen that the way to "do something to" a quantum state is with an operator. Thus, if a state is written in the form of a row or column vector, the way we do something to it is with an operator written in the form of a matrix.

I've emphasized in this chapter that one must choose a basis—whether formed by ordinary basis vectors or by quantum states—and work within that basis. As we will see, this is true for both states (vectors) and operators (matrices).

As a simplest first example, suppose $|\phi_1\rangle$ and $|\phi_2\rangle$ satisfy the eigenvalue equation $\hat{A}|\phi_j\rangle = a_j|\phi_j\rangle$, and form a basis, the *eigenbasis* of $\hat{A}$. Suppose we choose to work in this basis, and to let $\hat{A}$ act on its eigenstates. Then a 2×2 matrix with diagonal elements a_1 and a_2, and off-diagonal elements 0, will evidently do the job:

$$\hat{A}|\phi_1\rangle = a_1|\phi_1\rangle \longrightarrow \begin{pmatrix} a_1 & 0 \\ 0 & a_2 \end{pmatrix} \begin{pmatrix} 1 \\ 0 \end{pmatrix} = a_1 \begin{pmatrix} 1 \\ 0 \end{pmatrix}, \tag{6.13}$$

and similarly for $j = 2$. A matrix corresponding to an operator, such as that in Eq. (6.13), is often called the *matrix representation* of the operator.[4]

Far from applying only to the eigenstates of $\hat{A}$, though, the matrix in Eq. (6.13) is the correct matrix representation of $\hat{A}$ for *any* state written in the ϕ basis. To see this, consider a state $|\Psi\rangle$ that is an arbitrary superposition of $|\phi_1\rangle$ and $|\phi_2\rangle$, that is, $|\Psi\rangle = c_1|\phi_1\rangle + c_2|\phi_2\rangle$. The action of $\hat{A}$ on $|\Psi\rangle$ is:

$$\hat{A}|\Psi\rangle = \hat{A}\Big(c_1|\phi_1\rangle + c_2|\phi_2\rangle\Big) = a_1 c_1|\phi_1\rangle + a_2 c_2|\phi_2\rangle. \tag{6.14}$$

This is precisely the effect of the matrix of Eq. (6.13):

$$\begin{pmatrix} a_1 & 0 \\ 0 & a_2 \end{pmatrix} \begin{pmatrix} c_1 \\ c_2 \end{pmatrix} = \begin{pmatrix} a_1 c_1 \\ a_2 c_2 \end{pmatrix}. \tag{6.15}$$

6.3.2 Matrix Elements and Alternative Bases

I did not *derive* the correct matrix representation of $\hat{A}$: I simply wrote it down. Similarly, I could write $\hat{A}$ in its eigenbasis as

$$\hat{A} = \begin{pmatrix} \langle\phi_1|\hat{A}|\phi_1\rangle & \langle\phi_1|\hat{A}|\phi_2\rangle \\ \langle\phi_2|\hat{A}|\phi_1\rangle & \langle\phi_2|\hat{A}|\phi_2\rangle \end{pmatrix}, \tag{6.16}$$

where the only justification (for now) is that *it works*; that is, Eq. (6.16) reproduces the matrix of Eq. (6.15). The various $\langle\phi_j|\hat{A}|\phi_k\rangle$s are the *matrix elements* of $\hat{A}$ in the ϕ basis. Clearly, because we are working in the basis formed by the eigenstates of $\hat{A}$, the diagonal elements are the eigenvalues of $\hat{A}$, and the off-diagonal elements are 0.

Until now, the action of an operator on a state was dealt with by expanding the state in the operator's eigenbasis. In Chapter 5, however, I suggested that an operator could be written in a basis other than that formed by its eigenstates. To do so, the matrix elements are written in terms of the basis of choice. Suppose, for example, that the $|\chi\rangle$s are *not* eigenstates of $\hat{A}$. In the χ basis, then, $\hat{A}$ becomes:

$$\hat{A} = \begin{pmatrix} \langle\chi_1|\hat{A}|\chi_1\rangle & \langle\chi_1|\hat{A}|\chi_2\rangle \\ \langle\chi_2|\hat{A}|\chi_1\rangle & \langle\chi_2|\hat{A}|\chi_2\rangle \end{pmatrix}. \tag{6.17}$$

Note that this is the *same* operator as in Eqs. (6.15) and (6.16), but now expressed in the χ basis.

---

[4] Clearly, "representation" is used here in a different sense than in Chapter 4, where it referred to a particular basis.

In this basis the off-diagonal elements will *not*, in general, vanish. If, for example, $|\chi_1\rangle = \alpha_1|\phi_1\rangle + \alpha_2|\phi_2\rangle$, and $|\chi_2\rangle = \beta_1|\phi_1\rangle + \beta_2|\phi_2\rangle$, then the off-diagonal element A_{12} is

$$A_{12} = \langle\chi_1|\hat{A}|\chi_2\rangle = \left(\alpha_1^*\langle\phi_1| + \alpha_2^*\langle\phi_2|\right)\left(a_1\beta_1|\phi_1\rangle + a_2\beta_2|\phi_2\rangle\right)$$

$$= a_1\alpha_1^*\beta_1 + a_2\alpha_2^*\beta_2, \tag{6.18}$$

which in general does not vanish. It shouldn't be hard to also convince yourself that a diagonal element, such as $\langle\chi_1|\hat{A}|\chi_1\rangle$, will no longer simply be an eigenvalue of $\hat{A}$.

Suppose that one must deal with a state that is known in one basis, and an operator known in another. We may transform either the state or the operator to the other basis. One reason it may be convenient to choose the latter approach arises if we will need to repeatedly use our operator with different states. If we transform states, we will need to do so for each new state. But if we transform the operator, we need only do so once, for once we have obtained the operator in that basis, it works for all states written in that basis.

Many textbooks supply formal derivations that the matrix representation of an operator in an arbitrary basis takes the form of $\hat{A}$ in Eq. (6.17).[5] Instead, I will provide a qualitatively accurate explanation of why this is so, using $\hat{A}$ of Eq. (6.17) as an example.

Suppose $\hat{A}$ acts on a state $|\mu\rangle = \langle\chi_1|\mu\rangle|\chi_1\rangle + \langle\chi_2|\mu\rangle|\chi_2\rangle$ to form a new state $|\mu'\rangle$. Writing this out in explicit matrix form, we have

$$\begin{pmatrix} \langle\chi_1|\hat{A}|\chi_1\rangle & \langle\chi_1|\hat{A}|\chi_2\rangle \\ \langle\chi_2|\hat{A}|\chi_1\rangle & \langle\chi_2|\hat{A}|\chi_2\rangle \end{pmatrix} \begin{pmatrix} \langle\chi_1|\mu\rangle \\ \langle\chi_2|\mu\rangle \end{pmatrix} = \begin{pmatrix} \langle\chi_1|\mu'\rangle \\ \langle\chi_2|\mu'\rangle \end{pmatrix}. \tag{6.19}$$

Let's examine the $|\chi_1\rangle$ component of the new state $|\mu'\rangle$:

$$\langle\chi_1|\mu'\rangle = \langle\chi_1|\hat{A}|\chi_1\rangle\langle\chi_1|\mu\rangle + \langle\chi_1|\hat{A}|\chi_2\rangle\langle\chi_2|\mu\rangle. \tag{6.20}$$

What is this equation telling us?

Consider first the term $\langle\chi_1|\hat{A}|\chi_1\rangle\langle\chi_1|\mu\rangle$. Because $|\chi_1\rangle$ is *not* an eigenstate of $\hat{A}$, the state $\hat{A}|\chi_1\rangle$ must be a superposition of the basis states $|\chi_1\rangle$ and $|\chi_2\rangle$. The second factor, $\langle\chi_1|\mu\rangle$, is the projection of the state $|\mu\rangle$ onto the basis state $|\chi_1\rangle$: it's the "amount" of $|\chi_1\rangle$ in $|\mu\rangle$. Writing the first factor as $\langle\chi_1|(\hat{A}|\chi_1\rangle)$ suggests interpreting that factor as the projection of the state $\hat{A}|\chi_1\rangle$ onto the basis state $|\chi_1\rangle$. That is, it's the amount of $|\chi_1\rangle$ in $\hat{A}|\chi_1\rangle$.

In words, $\langle\chi_1|\hat{A}|\chi_1\rangle\langle\chi_1|\mu\rangle$ is the amount of $|\chi_1\rangle$ in the system state $|\mu\rangle$, times the amount of $|\chi_1\rangle$ in $\hat{A}|\chi_1\rangle$, the state created by the action of $\hat{A}$ on $|\chi_1\rangle$. Similarly, $\langle\chi_1|\hat{A}|\chi_2\rangle\langle\chi_2|\mu\rangle$ is the amount of $|\chi_2\rangle$ in the system state

---

[5] See, for example, *Liboff* (2003), Section 11.1.

$|\mu\rangle$, times the amount of $|\chi_1\rangle$ in $\hat{A}|\chi_2\rangle$, the state created by the action of $\hat{A}$ on $|\chi_2\rangle$. The sum of these is the total contribution of $|\chi_1\rangle$ to the new system state $|\mu'\rangle$, that is, it's $\langle\chi_1|\mu'\rangle$. An analogous argument applies to the $|\chi_2\rangle$ component of $|\mu'\rangle$.

6.3.3 Change of Basis

Changing representations is a fundamentally important quantum-mechanical operation. Moreover, by examining the matrix elements that effect such a transformation, we can better grasp the structure of operators when represented by matrices.

I'll denote as $\hat{S}$ a matrix that transforms a ket (a column vector) from the χ basis to the ϕ basis. Consider a state $|\mu\rangle = \langle\chi_1|\mu\rangle|\chi_1\rangle + \langle\chi_2|\mu\rangle|\chi_2\rangle = \langle\phi_1|\mu\rangle|\phi_1\rangle + \langle\phi_2|\mu\rangle|\phi_2\rangle$. The transformation matrix $\hat{S}$ must satisfy

$$\begin{pmatrix} S_{11} & S_{12} \\ S_{21} & S_{22} \end{pmatrix} \begin{pmatrix} \langle\chi_1|\mu\rangle \\ \langle\chi_2|\mu\rangle \end{pmatrix} = \begin{pmatrix} \langle\phi_1|\mu\rangle \\ \langle\phi_2|\mu\rangle \end{pmatrix}. \tag{6.21}$$

It turns out that $\hat{S}$ must take the form

$$\hat{S} = \begin{pmatrix} \langle\phi_1|\chi_1\rangle & \langle\phi_1|\chi_2\rangle \\ \langle\phi_2|\chi_1\rangle & \langle\phi_2|\chi_2\rangle \end{pmatrix}. \tag{6.22}$$

But why?

Let's write out one coefficient of the transformed state explicitly:

$$\langle\phi_1|\mu\rangle = \langle\phi_1|\chi_1\rangle\langle\chi_1|\mu\rangle + \langle\phi_1|\chi_2\rangle\langle\chi_2|\mu\rangle. \tag{6.23}$$

From Chapter 5 we know that the identity operator for this system may be written: $\hat{I} = |\chi_1\rangle\langle\chi_1| + |\chi_2\rangle\langle\chi_2|$. Thus, Eq. (6.23) is simply: $\langle\phi_1|\mu\rangle = \langle\phi_1|\hat{I}|\mu\rangle$. Nevertheless, it's somewhat instructive to dissect the terms in Eq. (6.23).

The term $\langle\phi_1|\chi_1\rangle\langle\chi_1|\mu\rangle$ is the projection of $|\mu\rangle$ onto $|\chi_1\rangle$ times the projection of $|\chi_1\rangle$ onto $|\phi_1\rangle$. Restated, it's the amount of $|\chi_1\rangle$ in $|\mu\rangle$ (the state we're transforming) times the amount of $|\phi_1\rangle$ in $|\chi_1\rangle$. Thus, this term gives the amount of $|\phi_1\rangle$ in $|\mu\rangle$ due to its $|\chi_1\rangle$ part. Similarly, $\langle\phi_1|\chi_2\rangle\langle\chi_2|\mu\rangle$ gives the amount of $|\phi_1\rangle$ in $|\mu\rangle$ due to its $|\chi_2\rangle$ part. The sum of these two contributions then gives the total amount of $|\phi_1\rangle$ in $|\mu\rangle$. An analogous argument holds for the other component of the transformed state, $\langle\phi_2|\mu\rangle$. So, $\hat{S}$ in Eq. (6.22) indeed transforms from the χ basis to the ϕ basis.

6.3.4 Adjoint, Hermitian, and Unitary Operators

Adjoint, Hermitian, and unitary operators assume particular forms in matrix mechanics. Moreover, matrix mechanics affords a means to make these rather abstract concepts more concrete.

By definition, the adjoint of an operator $\hat{B}$ is another operator $\hat{B}^\dagger$ that, acting to the left, yields the bra corresponding to the ket created by $\hat{B}$ acting to the right (cf. Chapter 5). That is, if $\hat{B}|\psi\rangle = |\psi'\rangle$, then $\langle\psi|\hat{B}^\dagger = \langle\psi'|$. In matrix mechanics, an adjoint operator, such as $\hat{B}^\dagger$, is the complex conjugate of the transpose of the original operator, $\hat{B}$. That is, $B^\dagger_{jk} = B^*_{kj}$. Thus, if:

$$\hat{B} = \begin{pmatrix} B_{11} & B_{12} \\ B_{21} & B_{22} \end{pmatrix}, \quad \text{then} \quad \hat{B}^\dagger = \begin{pmatrix} B^*_{11} & B^*_{21} \\ B^*_{12} & B^*_{22} \end{pmatrix}. \tag{6.24}$$

To see why this condition indeed yields the adjoint of an operator, I first need to clarify the phrase 'acting to the left'. Though commonly used, this terminology is a bit misleading. In matrix mechanics, saying that $\langle\psi|\hat{B}^\dagger$ denotes $\hat{B}^\dagger$ acting to the left on $\langle\psi|$ really means that the *row* vector $\langle\psi|$ multiplies the matrix $\hat{B}^\dagger$ in the usual way, acting to the *right*. Thus, $\langle\psi|\hat{B}^\dagger$ is

$$\langle\psi|\hat{B}^\dagger = (c^*_1, \ c^*_2) \begin{pmatrix} B^*_{11} & B^*_{21} \\ B^*_{12} & B^*_{22} \end{pmatrix}$$
$$= (c^*_1 B^*_{11} + c^*_2 B^*_{12}, \ c^*_1 B^*_{21} + c^*_2 B^*_{22}) = \langle\psi'|, \tag{6.25}$$

while $\hat{B}|\psi\rangle$ is

$$\hat{B}|\psi\rangle = \begin{pmatrix} B_{11} & B_{12} \\ B_{21} & B_{22} \end{pmatrix} \begin{pmatrix} c_1 \\ c_2 \end{pmatrix} = \begin{pmatrix} B_{11}c_1 + B_{12}c_2 \\ B_{21}c_1 + B_{22}c_2 \end{pmatrix} = |\psi'\rangle, \tag{6.26}$$

which shows that the conjugate transpose of an operator does indeed yield the adjoint.

Since a Hermitian operator is self-adjoint, its manifestation in matrix mechanics is now obvious. If $\hat{B}$ is Hermitian, then $\hat{B} = \hat{B}^\dagger$, so the matrix elements must obey:

$$B_{jk} = B^\dagger_{jk} = B^*_{kj}. \tag{6.27}$$

That is, a Hermitian operator must equal its transpose conjugate (its adjoint).

The defining characteristic of a unitary operator $\hat{U}$ is that its adjoint equals its inverse: $\hat{U}^\dagger = \hat{U}^{-1}$. Thus $\hat{U}^\dagger\hat{U} = \hat{U}^{-1}\hat{U} = \hat{I}$. An example is an operator that effects a change of basis, such as $\hat{S}$ in Eq. (6.22). One can easily check that $\hat{S}$ satisfies the unitarity condition.

In Section 6.3.3 we saw how the matrix $\hat{S}$ transforms a ket from the χ basis to the ϕ basis. As pointed out in Chapter 5, however, we can also transform the basis of an *operator*, via a unitary transformation (more explicitly, a *unitary-similarity transformation*). If $\hat{G}_\chi$ is an operator in the

χ basis, then $\hat{G}_\phi$, the *same* operator in the ϕ basis, is

$$\hat{G}_\phi = \hat{S}^{-1}\hat{G}_\chi\hat{S} = \hat{S}^\dagger\hat{G}_\chi\hat{S}. \tag{6.28}$$

The last equality holds, of course, because $\hat{S}$ is unitary.

6.4 Eigenvalue Equations

Consider a matrix mechanics eigenvalue equation for the simplest case of a two-state basis

$$\begin{pmatrix} A_{11} & A_{12} \\ A_{21} & A_{22} \end{pmatrix} \begin{pmatrix} c_1 \\ c_2 \end{pmatrix} = a \begin{pmatrix} c_1 \\ c_2 \end{pmatrix}. \tag{6.29}$$

Note that Eq. (6.29) does indeed fulfill the definition of an eigenvalue equation—an operator acting on a state must yield the same state, times a constant. The values of a that satisfy Eq. (6.29) are the eigenvalues. If A_{12} and/or A_{21}, as well as both c_1 and c_2, are non-zero, then we evidently are *not* working in the eigenbasis of $\hat{A}$.

Suppose we transform the state, and $\hat{A}$, into $\hat{A}$'s eigenbasis. Denoting the transformed matrix elements by primes, Eq. (6.29) takes the form

$$\begin{pmatrix} A'_{11} & 0 \\ 0 & A'_{22} \end{pmatrix} \begin{pmatrix} 1 \\ 0 \end{pmatrix} = \begin{pmatrix} a_1 & 0 \\ 0 & a_2 \end{pmatrix} \begin{pmatrix} 1 \\ 0 \end{pmatrix} = a_1 \begin{pmatrix} 1 \\ 0 \end{pmatrix}, \tag{6.30}$$

for the eigenstate with eigenvalue a_1, and similarly for the eigenstate corresponding to a_2. Appropriately, transforming to an operator's eigenbasis is often called *diagonalizing* the operator. Clearly, if we can transform our eigenvalue equation into the eigenbasis of $\hat{A}$—if we can diagonalize it—then we can simply read off the eigenvalues.

An algebraic route to solution of the eigenvalue problem in matrix mechanics is also available. An equation such as Eq. (6.29) is really a set (in this case, two) of *algebraic* equations

$$\begin{pmatrix} A_{11} - a & A_{12} \\ A_{21} & A_{22} - a \end{pmatrix} \begin{pmatrix} c_1 \\ c_2 \end{pmatrix} = \begin{pmatrix} 0 \\ 0 \end{pmatrix}. \tag{6.31}$$

From algebra, we know that solutions to such a set of equations are obtained by setting the corresponding determinant to zero:

$$\begin{vmatrix} A_{11} - a & A_{12} \\ A_{21} & A_{22} - a \end{vmatrix} = 0. \tag{6.32}$$

This yields a quadratic equation which is to be solved for the eigenvalues. For this two-state basis, this is a simple algebraic problem, with at most two eigenvalues (a_1 and a_2). Complexity increases very rapidly, however, as the number of basis states increases.

By focusing on two-state bases, our discussion in this chapter has been relatively simple and concrete. Although calculations may become very complicated for higher-dimensional bases, the fundamental ideas developed herein remain valid, and widely applicable.

6.5   Problems

1. Consider an operator $\hat{\Gamma} = \exp(-i\hat{G}\gamma/\hbar)$, where γ is a parameter. Assume that $\hat{G}|\phi_k\rangle = g_k|\phi_k\rangle$, and that k runs from 1 to 2.

   Use the step-by-step instructions below to calculate, as far as is possible, the matrix element Γ_{12} in the β basis, defined by $\hat{B}|\beta_n\rangle = b_n|\beta_n\rangle$ (assume that $\hat{B}$ and $\hat{G}$ do *not* share eigenstates).

   (i) Write out Γ_{12} in the β basis in bra-ket notation. Don't do any calculations yet.

   (ii) Construct an appropriate identity operator $\hat{I}$ (a "sum of projectors") that will serve to convert the states in the inner product to a basis that will facilitate calculation.

   (iii) Insert your $\hat{I}$ from Part (ii) into the expression of Part (i) as necessary to carry out the calculation. (Hint: you need to insert $\hat{I}$ twice.)

   (iv) Carry out the calculation as far as possible with the information given.

2. For a spin 1/2 system: $\hat{S}_\alpha|\pm\alpha\rangle = \pm\frac{\hbar}{2}|\pm\alpha\rangle$, where α can be x, y, or z. Note that the y and z bases are related by: $|\pm y\rangle = 1/\sqrt{2}|+z\rangle \pm i/\sqrt{2}|-z\rangle$.

   (i) The operator $\hat{S}_y$, in matrix representation, and in the z basis, is

   $$\hat{S}_y \longrightarrow \begin{pmatrix} 0 & -i\hbar/2 \\ \cdots & 0 \end{pmatrix},$$

   where I have not supplied the matrix element $S_{y(21)}$. (The subscripts 1 and 2 correspond to $+$ and $-$, respectively.) Calculate $S_{y(21)}$. Be explicit, so what you are doing is clear.

   (ii) State (in words) what kind of equation you should get if you carry out the matrix-mechanical operation $\hat{S}_y|+y\rangle$ using your answer to Part (i). Then carry out the calculation, to show that you get this.

3. This *mega-problem* is not terribly difficult, but it does have many pieces.

   Consider again a two-state basis. One set of basis states is comprised of $|\alpha_+\rangle$ and $|\alpha_-\rangle$; another of $|\beta_+\rangle$ and $|\beta_-\rangle$. The basis sets are related by:

   $$|\beta_+\rangle = \frac{1}{\sqrt{2}}|\alpha_+\rangle + \frac{i}{\sqrt{2}}|\alpha_-\rangle$$

   $$|\beta_-\rangle = \frac{1}{\sqrt{2}}|\alpha_+\rangle - \frac{i}{\sqrt{2}}|\alpha_-\rangle$$

You're interested in this system because your experimentalist colleagues have obtained some puzzling preliminary results. They'll need to run many more experiments, and calculate the predicted results for many different initial states.

Using matrix mechanics, and working in the α basis, do the following.

(i) Write out $|\alpha_+\rangle$ and $|\alpha_-\rangle$.

(ii) Write out $|\beta_+\rangle$ and $|\beta_-\rangle$.

(iii) Write out $\langle\beta_+|$ and $\langle\beta_-|$.

(iv) Writing matrix elements as inner products, construct a matrix $\hat{S}_{\alpha\beta}$ that transforms states from the α basis to the β basis. (Regard the $+$ and $-$ states as corresponding to element subscripts 1 and 2, respectively. Thus, for example, S_{+-} occupies the 1,2 position in the matrix.)

(v) Evaluate the elements of $\hat{S}_{\alpha\beta}$, and rewrite $\hat{S}_{\alpha\beta}$ using these matrix elements.

(vi) A change of basis represents a rotation in Hilbert space, and this corresponds to a unitary transformation. Use the defining condition for unitarity to determine whether $\hat{S}_{\alpha\beta}$ is unitary.

(vii) Show that $\hat{S}_{\alpha\beta}$ either is or is not Hermitian.

(viii) The experimentalists can prepare states that are known in the α basis. The eigenstates of the observable that they're interested in, though, are the $|\beta\rangle$s. The operator corresponding to this observable is $\hat{\beta}$, which satisfies

$$\hat{\beta}|\beta_\pm\rangle = \pm\kappa|\beta_\pm\rangle,$$

where κ is a constant. Write down the matrix $\hat{\beta}$, in the β basis.

(ix) Use matrix mechanics to calculate $\langle\psi|\hat{\beta}|\psi\rangle$, the expectation value of β, where $|\psi\rangle$ is an arbitrary state. From Part (viii), $|\psi\rangle$ is known in the α basis; but $\hat{\beta}$ is known in its eigenbasis. To calculate $\langle\beta\rangle$, transform the state(s) to the β basis.

(x) As a conscientious researcher, you always try to verify that your result makes sense. Check that your result in Part (ix) is the sort of mathematical object that must be obtained for the expectation value of a Hermitian operator.

(xi) Your colleagues are hoping to publish their results, with you as a co-author, in a prestigious journal, such as *Physical Review Letters*. They're depending on you to get things right. (It's one thing to be wrong; another to be wrong, in print, about a simple calculation.) To check your own work, you also calculate $\langle\psi|\hat{\beta}|\psi\rangle$ by another method, in which you transform the operator rather than the state. Carry out this transformation. Is the resulting operator Hermitian? Is this what you expect?

(xii) Calculate $\langle\psi|\hat{\beta}|\psi\rangle$ using the transformed operator of Part (xi).

(xiii) Check that the results of Parts (ix) and (xii) are the same. Argue *on physical grounds* that they must be. (Hint: what does $\langle \psi | \hat{\beta} | \psi \rangle$ represent physically?)

You now have a prediction for $\langle \beta \rangle$ in *any* state. Your experimentalist colleagues can compare your predictions with their measurement results.

4. Suppose the states $|\phi_1\rangle$ and $|\phi_2\rangle$ form a basis.

(i) Working in the ϕ basis, construct a 2×2 matrix, denoted $\hat{P}_2$, whose action on any state is to project out the $|\phi_2\rangle$ part.

(ii) Write out a matrix equation that shows explicitly that $\hat{P}_2$ works correctly on an arbitrary state. Then write out the corresponding equation in bra-ket notation. Explain clearly, in words, why your matrix $\hat{P}_2$ does what you want.

5. The states $|+z\rangle$ and $|-z\rangle$ form the z basis of a spin 1/2 system. Working in this basis, explicitly construct the matrix representations of $\hat{P}_+$ and $\hat{P}_-$, the projection operators onto the states $|+z\rangle$ and $|-z\rangle$, respectively. Then check your results by letting each matrix act on an arbitrary spin 1/2 state.

6. The Hamiltonian operator, $\hat{H}$, corresponds to energy. For some quantum system, the matrix representation of $\hat{H}$ is

$$\begin{pmatrix} \alpha_1 & 0 & 0 \\ 0 & \alpha_2 & 0 \\ 0 & 0 & \alpha_3 \end{pmatrix}, \tag{6.33}$$

where $\alpha_j = j^2 \hbar^2$. (For dimensional consistency, you may assume j^2 has dimensions of $[mass \times length^2]^{-1}$.)

(i) What are the possible measured values of the energy for this system?

(ii) Suppose the system is prepared in the state $|\Psi\rangle = c_1|E_1\rangle + c_2|E_2\rangle + c_3|E_3\rangle$, where the $|E_k\rangle$s are the eigenstates of $\hat{H}$. Suppose, further, that

$$\langle c_j | \Psi \rangle = \sqrt{\tfrac{i}{6}} \, .$$

What are the probabilities for obtaining each possible energy value?

(iii) An operator $\hat{A}$ satisfies: $\hat{A}|a_i\rangle = a_i|a_i\rangle$. Suppose that, in the A basis, (i.e., the basis formed by the $|a_n\rangle$s):

$$|E_1\rangle \longrightarrow \frac{1}{\sqrt{6}} \begin{pmatrix} 1 \\ \sqrt{2}\,i \\ \sqrt{3} \end{pmatrix} .$$

(a) What are $\langle A_1|E_1\rangle$, $\langle A_2|E_1\rangle$, and $\langle A_3|E_1\rangle$?

(b) What is the matrix element A_{11} in the A basis?

(c) What is A_{11} in the energy representation (basis)?

7. Suppose that for some system, $\hat{G}|\gamma_j\rangle = \gamma_j|\gamma_j\rangle$, where $j = 1, 2$.

(i) Construct the matrix representation of $\hat{G}$ in the γ basis.

(ii) Show explicitly that the matrix of Part (i) satisfies the above eigenvalue equation.

(iii) Suppose that the γ and β bases are related by:

$$|\gamma_1\rangle = \tfrac{1}{\sqrt{3}}|\beta_1\rangle - \sqrt{\tfrac{2}{3}}|\beta_2\rangle, \qquad |\gamma_2\rangle = \sqrt{\tfrac{2}{3}}|\beta_1\rangle + \tfrac{1}{\sqrt{3}}|\beta_2\rangle.$$

Construct the matrix representation of $\hat{G}$ in the β basis.

(iv) Show explicitly that if $\hat{G}$ of Part (iii) acts on one of its eigenstates, we get what we should.

7

Commutators and Uncertainty Relations

Thus, the more precisely the position is determined, the less precisely the momentum is known, and conversely. In this circumstance we see a direct physical interpretation of the equation $pq - qp = -i\hbar$.

I believe that one can fruitfully formulate the origin of the classical "orbit" in this way: the "orbit" comes into being only when we observe it.

As the statistical character of quantum theory is so closely linked to the inexactness of all perceptions, one might be led to the presumption that behind the perceived statistical world there still hides a "real" world in which causality holds... One can express the true state of affairs better in this way: Because all experiments are subject to the laws of quantum mechanics... it follows that quantum mechanics establishes the final failure of causality.

Werner Heisenberg[1]

Although I introduced the statistical interpretation of quantum mechanics in Chapter 3, the subsequent discussion has been rather formal—physical meaning seems to have faded from view. In this chapter I continue the formal development of quantum mechanics, but I also return to its physical interpretation.

A few basic features seem to be responsible for many of the puzzles of quantum mechanics. This chapter focuses on two such features: commutators and uncertainty relations. These topics are closely entwined, and both have been the subject of much debate regarding their import and interpretation. The introductory quotations—from Werner Heisenberg's original 1927 paper on the uncertainty relations—illustrate the profound physical and philosophical conclusions that he drew from those relations in quantum mechanics' infancy. The uncertainty relations[2] were historically viewed as fundamentally important—even as forming the basis for quantum mechanics itself.[3] Although this view has been largely eclipsed,

---

[1] Heisenberg (1927).

[2] The uncertainty relations are also known as the uncertainty principle, the indeterminacy relations, or the indeterminacy principle.

[3] See, for example, Jammer (1974), Chapter 3.

the uncertainty relations remain an important element in the structure of quantum mechanics.

I first develop the commutator, upon which the general form of the uncertainty relations is built. I then discuss the uncertainty relations themselves, including the interpretation of them that I regard as proper, and that is in harmony with the statistical interpretation of quantum mechanics.

More than usual, the discussion will take on historical overtones. The history of quantum mechanics is deeply fascinating, and deserving of study in its own right. My motivation, though, is more pragmatic: few topics in quantum mechanics have been subject to so many different and often fiercely defended interpretations as have the uncertainty relations. In particular, discussions found in early works may seem unclear, and unrelated to—perhaps even irreconcilable with—the modern approach that I adopt. Thus, it is worth contrasting this approach with other important perspectives that have been advanced since Heisenberg first introduced the uncertainty relations in 1927.

7.1 The Commutator

7.1.1 Definition and Characteristics

Before one can carry out calculations within some mathematical framework, the fundamental objects that comprise that framework must be defined, and the rules for manipulating those objects established. In elementary algebra, for example, we deal with numbers and functions. Two fundamental algebraic rules concern the multiplication of numbers and of functions. For two numbers a and b, and for two functions $f(x)$ and $g(x)$, multiplication is commutative:

$$ab = ba, \quad \text{and} \quad f(x)g(x) = g(x)f(x), \tag{7.1}$$

that is, the order of multiplication doesn't matter. With a bit of practice, the commutativity of both numbers and operators with respect to multiplication becomes "obvious", and one applies the rule without giving it any thought.

In quantum mechanics, the objects that we deal with are numbers, states, and operators. Perhaps the most fundamental rule in quantum mechanics (apart from the fact that the ordinary rules of algebra apply to numbers and functions) is that operators are linear. This means that an operator, say $\hat{G}$, obeys:

$$\hat{G}\big(a|\psi\rangle + b|\phi\rangle\big) = a\hat{G}|\psi\rangle + b\hat{G}|\phi\rangle. \tag{7.2}$$

It would be nice if we also had at hand a simple, universal rule for the "multiplication" of operators, that is, for the manipulation of expressions such as $\hat{A}\hat{B}$ relative to $\hat{B}\hat{A}$. Alas, no such rule exists.

To characterize the commutativity, or lack thereof, of two objects α and β, we define their *commutator*, $[\alpha, \beta]$, as

$$[\alpha, \beta] \equiv \alpha\beta - \beta\alpha. \tag{7.3}$$

From Eq. (7.1), if α and β are numbers, or functions, their commutator vanishes: they *commute*. The situation is far more subtle for two operators, say $\hat{F}$ and $\hat{G}$. The lack of a universal rule as to the commutativity of operators is manifested in the fact that whether $\hat{F}$ and $\hat{G}$ commute, that is, whether $[\hat{F}, \hat{G}] = 0$, depends on the operators in question.

In introductory treatments, the only commutator discussed may be that between the one-dimensional position and momentum operators (in position representation), $\hat{x} = x$ and $\hat{p}_x = -i\hbar \, d/dx$, respectively. Specifically

$$[\hat{x}, \hat{p}_x] = \hat{x}\hat{p}_x - \hat{p}_x\hat{x} = i\hbar. \tag{7.4}$$

This limitation is in part understandable: position and momentum are the observables that seem most intuitively palatable. Moreover, they were the first that the founders of quantum mechanics had to deal with, and their non-commutativity—the "$qp - pq$ swindle," as Schrödinger called it—was a great puzzlement.

But in implicitly setting $[\hat{x}, \hat{p}_x]$ as the prototypical commutator, a serious misconception may be introduced. That misconception stems from a fundamental characteristic of commutators.

Consider $[\hat{F}, \hat{G}] = \hat{F}\hat{G} - \hat{G}\hat{F}$. What *are* the objects $\hat{F}\hat{G}$ and $\hat{G}\hat{F}$? Consider $\hat{F}\hat{G}|\psi\rangle$. This expression says: first carry out the instructions specified by $\hat{G}$ (on $|\psi\rangle$), and then carry out those specified by $\hat{F}$. But we could combine these into a single instruction set, a new operator, say $\hat{K}_+ \equiv \hat{F}\hat{G}$. Similarly, $\hat{G}\hat{F}$ is also an operator, say $\hat{K}_-$. Clearly, then, $[\hat{F}, \hat{G}]$ is also an operator, here denoted $\hat{\kappa}$: $[\hat{F}, \hat{G}] = \hat{K}_+ - \hat{K}_- \equiv \hat{\kappa}$. Thus, *the commutator of two operators is, in general, another operator*. The commutator of Eq. (7.4)—a constant—is, therefore, an exceptional case.

When we write $[\hat{F}, \hat{G}] = \hat{F}\hat{G} - \hat{G}\hat{F}$, we've "evaluated" the commutator. But to evaluate its role *in a calculation*, we generally need more: we need to know the state it acts on. Suppose, for example, that $[\hat{F}, \hat{G}] = d^3/dx^3$; clearly we cannot evaluate d^3/dx^3 in a calculation without specifying the state (i.e., function) it acts on. So, a corollary to the fact that the commutator of two operators generally yields another operator is that *evaluation of a commutator in a calculation generally requires specification of the state the commutator acts on*.

There are, however, exceptions: Eq. (7.4) is one, another is $[d/dx, d^2/dx^2]$, which trivially vanishes. Moreover, it's not hard to show that, if $\hat{F}$ and $\hat{G}$ share a complete set of eigenstates, $[\hat{F}, \hat{G}]$ vanishes identically (i.e., for *any* state).

The possibilities for a commutator of two operators, then, are

- the commutator vanishes identically,
- the commutator is a constant,
- the commutator is an operator; a state must in general be specified to evaluate the commutator in a calculation.[4]

One practical result of the preceding discussion is that if two operators share a complete set of eigenstates, they commute. In that case, we are free to change the order of operation of two such operators in a calculation, regardless of the state the operators may act upon.

The commutator also enters quantum mechanics in other, more fundamental ways. One is that the commutator of two operators is related to symmetries, conservation principles, and the group-theoretical structure of quantum mechanics. Because these are rather advanced topics, we will largely ignore them, apart from a brief discussion in Appendix D.

Two applications of the commutator play important roles even in elementary quantum mechanics. First, the time dependence of the expectation value of some operator $\hat{G}$ in the state $|\psi(t)\rangle$ is given by

$$\frac{d}{dt}\langle\psi(t)|\hat{G}|\psi(t)\rangle = \frac{i}{\hbar}\langle\psi(t)|[\hat{H},\hat{G}]|\psi(t)\rangle + \langle\psi(t)|\frac{\partial\hat{G}}{\partial t}|\psi(t)\rangle, \qquad (7.5)$$

where $\hat{H}$ is the Hamiltonian operator. I state this here only to point out that the commutator plays an important role in time-dependent quantum mechanics.[5] Second, the general formulation of the uncertainty relations, discussed in Section 7.2, is based on the commutator.

7.1.2 Commutators in Matrix Mechanics

Before leaving the commutator proper, I illustrate its manifestation in matrix mechanics. For a two-state basis, an operator is a 2×2 matrix. It's easy to show that matrix multiplication is, in general, not commutative. If $\hat{F}$ and $\hat{G}$ are operators, then $[\hat{F},\hat{G}]$ is

$$\begin{pmatrix} F_{11} & F_{12} \\ F_{21} & F_{22} \end{pmatrix}\begin{pmatrix} G_{11} & G_{12} \\ G_{21} & G_{22} \end{pmatrix} - \begin{pmatrix} G_{11} & G_{12} \\ G_{21} & G_{22} \end{pmatrix}\begin{pmatrix} F_{11} & F_{12} \\ F_{21} & F_{22} \end{pmatrix} \qquad (7.6)$$

In general, none of the matrix elements of $[\hat{F},\hat{G}]$ vanish. For example, the 1,1 element is

$$F_{12}G_{21} - G_{12}F_{21}. \qquad (7.7)$$

Note that the action of the matrix resulting from Eq. (7.6) is, in general, state dependent. An easy way to see this is to write out the matrix $[\hat{F},\hat{G}]$

---

[4] Examples are the commutators of the angular momentum operators. See Chapter 8.
[5] Time dependence is discussed in Chapter 11. Equation (7.5) is derived in Section 11.3.

explicitly, and allow it to act (separately) on the two basis states of the (unspecified) basis in which we are working.

We could, however, imagine a case such that

$$\hat{F}\hat{G} - \hat{G}\hat{F} = \begin{pmatrix} \alpha & 0 \\ 0 & \alpha \end{pmatrix}, \tag{7.8}$$

with α a constant. Then the action of $[\hat{F}, \hat{G}]$ on *any* state is simply multiplication of each component by α—the action of $[\hat{F}, \hat{G}]$ is state *independent*. Thus, Eq. (7.8), like Eq. (7.4), is a special case.

Finally, let's consider a case where $[\hat{F}, \hat{G}] = 0$ identically. We know that this holds if $\hat{F}$ and $\hat{G}$ share a complete set of eigenstates. Writing $[\hat{F}, \hat{G}]$ in the two operators' mutual eigenbasis, we have

$$\begin{pmatrix} F_{11} & 0 \\ 0 & F_{22} \end{pmatrix} \begin{pmatrix} G_{11} & 0 \\ 0 & G_{22} \end{pmatrix} - \begin{pmatrix} G_{11} & 0 \\ 0 & G_{22} \end{pmatrix} \begin{pmatrix} F_{11} & 0 \\ 0 & F_{22} \end{pmatrix}$$

$$= \begin{pmatrix} 0 & 0 \\ 0 & 0 \end{pmatrix}, \tag{7.9}$$

whose action on any state is obvious.

7.2 The Uncertainty Relations

7.2.1 Uncertainty Products

In Chapter 3 we saw that the term "uncertainty" may refer either to the uncertainty in a set of measurement results, or to the uncertainty in a theoretical (calculated) quantity. In this section we will be concerned with the latter case—specifically, the uncertainty in the probability distributions that arise in quantum mechanics. The fundamental importance that attaches to the uncertainty relations, however, stems not from the uncertainty in any single observable, but in the *product* of uncertainties of *different* observables.

The uncertainty in a probability distribution describes a characteristic of that distribution—roughly speaking, its width, or spread. Suppose that the probability distributions for two quantities are not independent—that they are constrained such that both distributions cannot be made arbitrarily narrow. It could be the case, for example, that if one distribution is narrow, then the other must be at least wide enough that the product of the two uncertainties obeys some lower bound. This is the essence of the quantum-mechanical uncertainty relations.

Just as the first exposure to commutators is often through $[\hat{x}, \hat{p}_x] = i\hbar$, so the first exposure to the uncertainty relations is often through $\Delta x \Delta p_x \geq \hbar/2$.[6] This inequality, a constraint on the product of two uncertainties,

---

[6] In quantum mechanics, the standard deviation (uncertainty) in some quantity y is typically not denoted $\sigma(y)$, as in Chapter 3, but Δy.

implies that the probability distributions for x and p_x are somehow related. In fact we know that this is true: a well known mathematical technique, the Fourier transform, takes a state from the position representation to the momentum representation, and vice versa. Thus, the x representation of a state (wavefunction) fully determines the state's p_x representation. This is a particular case of a central theme of Chapter 4: that there is but *one* system state and, by an appropriate change of basis, we can obtain the probability distribution for *any* relevant observable.

7.2.2 General Form of the Uncertainty Relations

In Section 7.1.1 we found it unfortunate that $[\hat{x}, \hat{p}_x] = i\hbar$ often plays the role of prototypical commutator. So, too, it is unfortunate—and for similar reasons—that $\Delta x \Delta p_x \geq \hbar/2$ often plays the role of prototypical uncertainty relation. For Heisenberg, the profound questions that swirled about the physical meaning of position and momentum in quantum theory in 1927 (and that largely survive today) found focus in $\Delta x \Delta p_x \geq \hbar/2$. But this uncertainty relation obscures fundamental characteristics of uncertainty relations generally.

It was some two years after Heisenberg's uncertainty paper appeared that H. P. Robertson presented what is now the best-known general form of the uncertainty relations.[7] In this general formulation, position and momentum play no privileged role. For any two Hermitian operators $\hat{A}$ and $\hat{B}$, Robertson found that

$$\Delta A \Delta B \geq \frac{|\langle\, i[\hat{A}, \hat{B}]\, \rangle|}{2} \tag{7.10}$$

Derivations of this general form are readily available in textbooks. Our interest is in elucidating its facets and interpretation.[8]

Because $\Delta A \Delta B$ is so closely related to $\langle[\hat{A}, \hat{B}]\rangle$, we expect that uncertainty products and commutators will share important characteristics. Commutators generally result in new operators, so that calculations involving commutators are generally *state dependent*. Thus uncertainty products, also, are generally state dependent—a fact made very clear by writing out the expectation value $\langle[\hat{A}, \hat{B}]\rangle$ in Eq. (7.10) as $\langle\Psi|[\hat{A}, \hat{B}]|\Psi\rangle$. Clearly, $\Delta x \Delta p_x \geq \hbar/2$ is an exceptional case (as is $[\hat{x}, \hat{p}_x] = i\hbar$ for commutators).

As we saw in Section 7.1, if two operators share a complete set of eigenstates, their commutator vanishes identically. Evidently, then, the uncertainty product in the observables corresponding to such operators must also vanish identically, that is, for any state.

There is thus a close correspondence between the characteristics of commutators and those of uncertainty relations. Remember, however, that the

---

[7] Robertson (1929).

[8] Time is a parameter in quantum mechanics, not a Hermitian operator. Because of this, the so-called "energy-time uncertainty relation" (see Chapter 11) cannot be obtained from Eq. (7.10)—a statement about Hermitian operators.

uncertainty relations are *in*equalities. If, say, $[\hat{F}, \hat{G}]$ vanishes identically, then clearly $\Delta F \Delta G \geq 0$. This does *not*, however, imply that either $\Delta F = 0$ or $\Delta G = 0$, since the uncertainty relation only sets a lower bound.

But can't we do more than determine a lower bound? Isn't the quantum world better defined than this? Yes. If we specify two operators and a state, then we can always, in principle, calculate an uncertainty *equality*: the actual product of the standard deviations of the two observables for the given state.

Consider the important case of x and p_x. For every possible state, corresponding x and p_x probability distributions exist. For each such pair of distributions, in turn, a perfectly well defined uncertainty product $\Delta x \Delta p_x$ exists. *Any* state must satisfy the uncertainty *relation* $\Delta x \Delta p_x \geq \hbar/2$, but the uncertainty *equality* formed by the lower limit, $\Delta x \Delta p_x = \hbar/2$, is satisfied only if the state is a Gaussian (a result derived in many textbooks).[9]

What, then, of the general case when the uncertainty product *is* state dependent? Specification of the state then determines both the uncertainty relation (the inequality) and the uncertainty equality. Although Eq. (7.10) only yields an inequality, it does provide a clear, concise statement about the constraints between quantum probability distributions; typically we need not calculate uncertainty equalities.

7.2.3 Interpretations

The uncertainty relations are not simply a mathematical result, derived in antiseptic fashion from the formal structure of quantum mechanics. They are also the vehicle by which Heisenberg and others tried to reconcile the concepts of classical physics with the new quantum mechanics.[10]

In this section I will first discuss two physical interpretations of the uncertainty relations that Heisenberg advanced. I do so not because they are compelling, but because they and their progeny have spawned so many varied and often conflicting statements. They thus form an important historical and conceptual backdrop for the approach that I advocate, and that we then turn to: interpreting the uncertainty relations in the context of the statistical interpretation of quantum mechanics.

Measurement interactions As Max Jammer has pointed out,[11] in formulating the uncertainty relations Heisenberg asked two distinct but related questions:

(1) Does the formalism allow for the fact that the position of a particle and its velocity are determinable at a given moment only

---

[9] That is, $\Delta x \Delta p_x \geq \hbar/2$ is state independent, but $\Delta x \Delta p_x = \hbar/2$ is state dependent.

[10] A highly recommended discussion of the uncertainty relations in mathematical, historical, and interpretive context is Hilgevoord (2001).

[11] Jammer (1974), Section 3.2.

with a limited degree of precision? (2) Would such imprecision, if admitted by the theory, be compatible with the optimum of accuracy obtainable in experimental measurements?

Clearly, (1) has to do with the formalism—the mathematical structure—of quantum mechanics, while (2) concerns the limitations of experiments. Heisenberg's answers to these questions differed accordingly.

To address (2), he considered a *gedanken*-experiment (thought experiment) involving a microscope in which an electron is illuminated with a gamma ray (a high-energy photon). Briefly, the argument is as follows. As the energy of the photon is increased, the position of the electron may be more accurately determined (thus the use of gamma rays); however, higher energy also results in a greater, uncontrollable change in momentum. The gamma-ray microscope is thus one manifestation of the opening quote of this chapter: "the more precisely the position is determined, the less precisely the momentum is known, and conversely."

Simple as it may seem, the implications of the gamma-ray microscope are far from obvious. Recall the provocative statement of Heisenberg and Max Born in Chapter 3: "We maintain that quantum mechanics is a complete theory; its basic physical and mathematical hypotheses are not further susceptible of modifications."

Yet the gamma-ray microscope does not support completeness—in fact, it contradicts it. In Heisenberg's microscope, perfectly well-defined values of the electron's position and momentum evidently *exist*; yet there is no experiment by which both can be perfectly *known*. And Heisenberg's experimentally-based argument is apparently reflected in the quantum formalism, through the uncertainty relations. Taken at face value, then, the gamma-ray microscope contradicts completeness, for if the electron's position and momentum are precisely defined, and exist, yet quantum mechanics cannot account for that fact, the theory must be *in*complete.[12]

The analysis of *gedanken*-experiments as an approach to the uncertainty relations is important for another reason: such experiments formed the basis for intense discussions between Neils Bohr and Albert Einstein regarding the interpretation of quantum mechanics—most notably at the 1927 Solvay congress. These were no ordinary academic debates. Jammer has called them "one of the great scientific debates in the history of physics . . . a clash between two of the greatest minds of their time."[13] As such, they rightfully occupy a special place in the history of modern physics. Perhaps because of them, expositions of the uncertainty relations have often been, and sometimes still are, framed in terms of *gedanken*-experiments. Philosophical arguments aside, however, the fact is that we do not need

---

[12] This isn't intended as proof of quantum mechanics' incompleteness, but it does suggest that this interpretation of the uncertainty relations is open to question.

[13] Jammer (1974), p. 120.

to invoke physical pictures such as those utilized in the gamma-ray micro-scope *gedanken*-experiment to acquire an operational understanding of the uncertainty relations.

Simultaneous measurements To answer question (1), Heisenberg took the Fourier transform of a Gaussian wavefunction, thus obtaining the state in momentum representation (see Chapter 12). Calculating the product of the standard deviations in x and p_x, he obtained the now famous uncertainty relation: $\Delta x \Delta p_x \geq \hbar/2$. How do we interpret this result?

The mathematical uncertainties, Δx and Δp_x, presumably correspond physically to a spread in measurement results. Because both Δx and Δp_x are calculated for the state at some particular time, we may interpret them as the spreads in x and p_x for *simultaneous* measurements. A Gaussian wavefunction is mathematically simple, but the fact is that, quite typi-cally, at any time t a wavefunction generally yields non-trivial probability distributions for both x and p_x.

Whether or not simultaneous measurements are in fact possible has been a matter of some debate. Regardless, if one insists on the complete-ness of quantum mechanics, things become outright weird. If quantum mechanics is complete, then the state must fully describe an individual system, say an electron. But all information about the electron must be contained in the state, and the state yields "blurry" distributions of x and p_x. Thus, the uncertainty relations evidently imply constraints on the *exis-tence* of simultaneously well-defined properties of a single system—that is, not only are the *probability distributions* for x and p_x indistinct, *so are x and p_x themselves*. Again, this strange result arises from an insistence on the completeness of quantum mechanics.

It is important to realize that even in this interpretation, one obtains an eigenvalue—a single, well-defined result—in any one measurement. If this were not true, then the third postulate (see Chapter 2) would be violated. Thus, the blurring of, say, x and p_x does not appear in any individual measurement, but only in an ensemble of measurements.[14]

Statistical interpretation If you're familiar with the difficult questions surrounding the measurement interaction and simultaneous measurement approaches to the uncertainty relations, the statistical approach may seem like a breath of fresh air. In Chapter 3 we suggested that the statistical interpretation of quantum mechanics instructs us to take seriously what the theory does tell us, and to not take seriously what it doesn't tell us.

What does an uncertainty relation, such as $\Delta x \Delta p_x \geq \hbar/2$, tell us? Clearly, it tells us that the probability distributions of x and p_x are not inde-pendent. Moreover, we expect probabilities to be reflected in an ensemble

---

[14] This arises from the notorious "collapse of the wavefunction", in which measure-ment suddenly and mysteriously collapses the state from a blurry distribution to an eigenstate.

of measurements. So Δx and Δp_x, the spreads in x and p_x, respectively, are manifested in a series of measurements on identically prepared sytems (e.g. electrons).

As in the simultaneous measurements interpretation, Δx and Δp_x apply to a state at a particular time. But in the statistical interpretation, probabilities refer to ensembles, so there is no reason to interpret uncertainties as necessarily applying to simultaneous measurements of both x and p_x on a single system. We interpret Δx and Δp_x as the spreads in x and p_x, respectively, that appear in making many separate measurements on identically prepared systems—one measurement of either x or p_x, *not both*, per system.

It may seem puzzling, but even in the simultaneous measurements approach, probabilities are always manifest in a series of measurements, never in a single measurement. That is, even if we somehow simultaneously measure x and p_x, we can obtain no evidence of a probability distribution until multiple measurements are made. The real difference between the simultaneous and statistical interpretations of the uncertainty relations lies not in what is calculated or measured, but in the meaning we attach to the results.

The measurement interaction interpretation is an attempt to understand the uncertainty relations in terms of largely classical physical pictures. The simultaneous measurements interpretation is more abstract and less classical, but arguably reads too much into the uncertainty relations. The statistical interpretation offers no physical explanation for the uncertainty relations, but simply implores us to interpret them operationally, as we would in the laboratory.

In Chapter 3, I argued that a chief virtue of the statistical interpretation of quantum mechanics is that it affords us a means to correctly think about quantum mechanics without becoming unnecessarily embroiled in foundational issues. Just that virtue is at work here. We take probabilities as the fundamental stuff of quantum mechanics, and that's enough to correctly represent how quantum mechanics is used in actual practice. And we let the experts argue about precisely what, if anything, those probabilities tell us about the world *beneath* the level of the formalism.

7.2.4 Reflections

The uncertainty relations reflect the fact that quantum-mechanical probability distributions are both fundamental and interdependent. But why is this so significant?

For a classical particle—the most basic object in classical mechanics—position and momentum are both independent and well defined. The particle's position does not determine its momentum (and vice versa), and both quantities may be specified to arbitrary precision. The uncertainty relations tell us, however, that quantum probability distributions *cannot* be independent. It is, for example, impossible for a quantum state to possess a finite probability distribution in x and no spread in p_x (i.e. $\Delta x < \infty$ and $\Delta p_x = 0$), or conversely, and similarly for many other pairs of observable quantities. Thus, quantum states are such that observable quantities *cannot*, in general, be specified to arbitrary precision. This is a remarkable result, for the quantum state (not a particle) is the most basic object in quantum mechanics—indeed, if one accepts completeness, it is the most basic object in the *world*.

The uncertainty relations are in fact deep statements of ideas that I have emphasized from early on: the fundamentally probabilistic nature of quantum mechanics, and the embodiment, within the quantum state, of *all* probabilities for the system. In a sense, however, they are simply a consequence of the superposition principle—the fact that we may expand a state in the eigenstates of different Hermitian operators, and from these expansions obtain the relevant probabilities. Different expansions yield different probability distributions, yet they cannot be independent, since all are extracted from the *same* state. The uncertainty relations quantify this fact.

The uncertainty relations have largely faded from the forefront of physical thought—not because they are incorrect or irrelevant, but because they are but one manifestation of the deep, binding thread woven throughout quantum mechanics by the superposition principle.

It is difficult now to imagine the intellectual fortitude that Heisenberg brought to bear in his pioneering work on uncertainty. Although our perspective on them has evolved greatly since 1927, the uncertainty relations remain one of the great intellectual constructs not only of 20th century physics, but of human thought generally. As John Rigden wrote,

> When first-rate minds are engaged in the intellectual activity called physics, as was the case in February 1927 when Heisenberg was struggling with the "$pq - qp$ swindle," it is an activity with no equivalent ... in any intellectual arena except, possibly, first-rate theological thinking. These special times in physics do not come often, but when they do, physicists must often create new constructs for which neither previous experience nor previous thought patterns provide guidance...Soon the new ideas become the basis for empirical predictions and, in the process, a "sense of understanding" emerges. However, in the end, the basic concepts of physics are aloof, they remain outside our ability to convey their meaning.

Rigden continues:

> With a bright, attentive person eagerly awaiting new insight, I do *not* believe I could convey the essence of the physics revolution ... With great erudition I could talk about electrons passing through a double slit, about wave-particle duality, about the inherent difficulties of knowing where an electron is and where

it is going. My best efforts, however, would fail: the essence
of quantum mechanics would remain mysterious, aloof from my
student's comprehension and alas, aloof from mine.[15]

Heisenberg, Einstein, Born and others ushered in the age of modern physics,
of relativity and quantum mechanics. But so too did they usher in an age in
which even physicists face a gap between physical prediction and physical
understanding—a gap between the world we perceive, and the world of
which modern physics speaks.

7.3 Problems

1. Suppose that, for some particular system and state, $\Delta D \Delta E \geq \Gamma$,
   where Γ is a positive number. Consider the following statement: *Physically, $\Delta D \Delta E \geq \Gamma$ means that a* **single** *measurement of D (or E)will*
   not *result in a sharp value for D (or E).*
   If you agree with this statement, say so, and explain how the statement
   is reflected in the above uncertainty relation. If you disagree, say so,
   and briefly explain what $\Delta D \Delta E \geq \Gamma$ *does* mean in terms of physical
   measurements.

2. The three Hermitian operators,

$$\hat{S}_x = \frac{\hbar}{2} \begin{pmatrix} 0 & 1 \\ 1 & 0 \end{pmatrix}, \quad \hat{S}_y = \frac{\hbar}{2} \begin{pmatrix} 0 & -i \\ i & 0 \end{pmatrix}, \quad \hat{S}_z = \frac{\hbar}{2} \begin{pmatrix} 1 & 0 \\ 0 & -1 \end{pmatrix},$$

represent the spin angular momentum for a spin $\frac{1}{2}$ system. The eigen-
states of $\hat{S}_x$ are denoted $|+x\rangle$ and $|-x\rangle$, and similarly for the
eigenstates of $\hat{S}_y$ and $\hat{S}_z$. The eigenstates of $\hat{S}_x$ and $\hat{S}_y$, written in
the $|\pm z\rangle$ basis, are

$$|\pm x\rangle = \frac{1}{\sqrt{2}}\Big(|+z\rangle \pm |-z\rangle\Big), \qquad |\pm y\rangle = \frac{1}{\sqrt{2}}\Big(|+z\rangle \pm i|-z\rangle\Big)$$

(i) The operators are all written in the same basis. What basis is it?
How do you know? (You should be able to answer this question simply
by inspection.)
(ii) Find all three (non-trivial) commutators of these operators.
(iii) Write your results explicitly in terms of $\hat{S}_x$, $\hat{S}_y$, and $\hat{S}_z$ (i.e., with
no matrices appearing).
(iv) Evaluate the uncertainty relations $\Delta S_x \Delta S_y$ and $\Delta S_y \Delta S_z$ for the
states:

$$|\psi_1\rangle = \begin{pmatrix} 0 \\ 1 \end{pmatrix}, \qquad |\psi_2\rangle = \frac{1}{\sqrt{2}} \begin{pmatrix} 1 \\ 1 \end{pmatrix}.$$

[15] Rigden (1987).

(The states are written in the same basis as the operators, above.)

(v) In Part (iv) we obtained:

(a) the uncertainty relation between S_x and S_y for the state $|\psi_1\rangle$,

(b) the uncertainty relation between S_x and S_y for the state $|\psi_2\rangle$,

(c) the uncertainty relation between S_y and S_z for the state $|\psi_1\rangle$,

(d) the uncertainty relation between S_y and S_z for the state $|\psi_2\rangle$.

For each case, argue that the uncertainty relation obtained in Part (iv) is consistent with what would be expected, based on the nature of the system state with respect to the relevant operators.

3. Show that: $\left[\hat{A}, [\hat{B}, \hat{C}]\right] - \left[\hat{B}[\hat{A}, \hat{C}]\right] = [\hat{A}, \hat{B}]\hat{C} + \hat{C}[\hat{B}, \hat{A}]$.

4. Although no basis was specified for Eq. (7.6), the two basis states (column vectors) are easily written down. Write out the matrix resulting from Eq. (7.6) explicitly, and allow it to act (separately) on these two basis states. Explain how this shows that the action of the matrix $[\hat{F}, \hat{G}]$ is state dependent.

8

Angular Momentum

Raffiniert ist der Herrgott, aber boshaft ist er nicht.
Albert Einstein[1]

"Subtle is the Lord," Einstein famously remarked, "but malicious He is not." Some years later, he explained: "Nature hides her secret because of her essential loftiness, but not by means of ruse." Nature—that is, physics—is, after all, largely susceptible to human comprehension, but it need not behave in accord with our naive hopes or expectations.

In this chapter, I first review angular momentum in classical mechanics, and then press on into quantum-mechanical angular momentum. In both cases, angular momentum is of fundamental importance. In quantum mechanics, however, angular momentum forces us to deal with the physical world on an even more abstract level than we have so far. But that is the nature of quantum mechanics, regardless of what we might wish the world to be. "If God created the world," said Einstein, "his primary concern was certainly not to make its understanding easy for us."[2]

8.1 Angular Momentum in Classical Mechanics

Both mathematically and conceptually, angular momentum in quantum mechanics bears little resemblance to its classical counterpart. Nevertheless, a brief review of the classical case is worthwhile, so as to refresh old ideas, and perhaps dispel misconceptions, before plunging into quantum angular momentum.

First, consider two vectors $\vec{A}$ and $\vec{B}$, and their *vector product*, denoted $\vec{A} \times \vec{B}$, and defined as[3]

$$\vec{A} \times \vec{B} \equiv (A_y B_z - A_z B_y)\hat{x} - (A_x B_z - A_z B_x)\hat{y} + (A_x B_y - A_y B_x)\hat{z} = \vec{C}.$$
$$(8.1)$$

Note that $\vec{A} \times \vec{B}$ yields another vector—thus the name *vector* product[4]—which we've called $\vec{C}$. It is a direct mathematical consequence of

---

[1] Said to Oswald Veblen in 1921. See Sayen (1985), pp. 50–51.
[2] Calaprice (2000), p. 218.
[3] In this section, the hat notation denotes a unit vector rather than an operator.
[4] $\vec{A} \times \vec{B}$ is also called the *cross product*. Compare $\vec{A} \cdot \vec{B}$, the *scalar*, or *dot*, product.

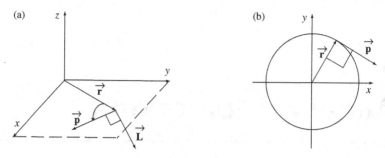

Fig. 8.1 (*a*) The classical angular momentum, $\vec{L}$, is perpendicular to both $\vec{r}$ and $\vec{p}$ (here $\vec{p}$ does *not* lie in the xy plane); (*b*) the simplified case of circular motion in the xy plane (here $\vec{r}$ and $\vec{p}$ are perpendicular; $\vec{L}$ points into the page).

Eq. (8.1) that $|\vec{A} \times \vec{B}| = AB \sin\theta$, where θ is the angle from $\vec{A}$ to $\vec{B}$. Moreover, it's easy to show that $(\vec{A} \times \vec{B}) \cdot \vec{A} = (\vec{A} \times \vec{B}) \cdot \vec{B} = 0$, so the direction of $\vec{C}$ must be perpendicular to the plane formed by $\vec{A}$ and $\vec{B}$. There are only two directions perpendicular to a plane, and the sole purpose of the famous "right-hand rule" is to (correctly) choose between those two possible directions.[5]

Returning to physics, the classical angular momentum, $\vec{L}$, is defined as

$$\vec{L} \equiv \vec{r} \times \vec{p} = (r_y p_z - r_z p_y)\hat{x} - (r_x p_z - r_z p_x)\hat{y} + (r_x p_y - r_y p_x)\hat{z}. \quad (8.2)$$

Here $\vec{p} = m\vec{v}$, the linear momentum, and $\vec{r}$ is the position vector from the point about which $\vec{L}$ is calculated to the object whose angular momentum we are calculating (see Fig. 8.1(*a*)).

Equation (8.2) is the general and complete definition of classical angular momentum: nothing need be added, and no restrictions need be imposed. This fact has important, and often overlooked, implications.

First, we can choose *any* point about which to calculate $\vec{L}$. If the object is rotating about some point, such as a wheel rotating on an axle, we *usually* choose that point of rotation as the point about which to calculate the angular momentum. (In both Figs. 8.1(*a*) and 8.1(*b*), $\vec{L}$ is calculated about the origin.) But the definition of $\vec{L}$, Eq. (8.2), does not *require* us to do so.

In addition, Eq. (8.2) places no restriction on the nature of the object's motion. The object *could*, for example, rotate in a circle, but it *need* not do so. Whether it rotates, moves in a straight line, or describes some complicated trajectory, $\vec{L}$ remains a well-defined quantity. Regardless of the point about which we choose to calculate $\vec{L}$, then, and regardless of the object's

---

[5] Discussion of the right-hand rule can be found in most any introductory physics text.

motion, $\vec{L}$ is a perfectly well-defined quantity. Whether $\vec{L}$ is of any interest for a particular case is, of course, a separate question.

Introductory discussions often focus on the magnitude of $\vec{L}$ (i.e., $L = rp\sin\theta$), along with the right-hand rule. It should be clear from our discussion of the vector product that this approach is simply a reformulation of Eq. (8.2).

Such discussions also often focus on circular motion, with $\vec{L}$ calculated about the axis of rotation (as in Fig. 8.1(b)). This simplified treatment makes pedagogical sense, but it can lead to misconceptions such as those outlined above. First, in this case $\vec{r}$ and $\vec{p}$ are *always* perpendicular, so $L = rp\sin\theta \rightarrow L = rp$. Also, $\vec{r}$ is taken from the axis of rotation, which can lead to the (false) assumption that angular momentum always deals with rotation about an axis, and that $\vec{L}$ must be calculated about that axis.

Finally, because the circular motion is confined to a plane, there are only two possible directions for $\vec{L}$ at all times: into the page or out of the page. Thus, we can represent $\vec{L}$'s direction with + and − signs alone. In Fig. 8.1(b), for example, $\vec{L}$ points into the page; we could assign this direction of $\vec{L}$ a − sign. For counter-clockwise rotation, $\vec{L}$ would point out of the page, and we could assign this direction a + sign. For this simplified case, with only two possible directions for $\vec{L}$, assigning direction through signs alone works fine. Still, it obscures the full vector nature of $\vec{L}$, and it can instill the misconception that angular momentum can always be treated this simply—which it clearly can't.[6]

8.2 Basics of Quantum Angular Momentum

8.2.1 Operators and Commutation Relations

Quantum-mechanical angular momentum is a subject both deep and wide, and our goal can only be to introduce the essentials. Section 8.1 served largely as a review of angular momentum in classical mechanics, but it also provided the tools to begin discussing quantum angular momentum.

In a real sense, the commutation relations for angular momentum operators form the foundation for quantum angular momentum generally. These commutators may be obtained through a mathematically sophisticated approach based on continuous transformations (see Appendix D). For our purposes, however, a less elegant but more straightforward approach is preferable.

From Eq. (8.2), we see that the z component of the classical angular momentum is: $L_z = (r_x p_y - r_y p_x)$. How could we construct $\hat{L}_z$, the quantum-mechanical operator corresponding to L_z? For a one-dimensional

---

[6] Similarly, for one-dimensional motion (e.g. along the x axis) there are only two possible directions, and these may be described with + and − signs rather than vector notation. Clearly, this fails for multi-dimensional motion.

system in position representation (i.e. $\psi = \psi(x, t)$), we know that the position operator is just x itself, $\hat{x} = x$, and the momentum operator is given by $\hat{p}_x = -i\hbar \frac{\partial}{\partial x}$. Exactly analogous expressions hold for $\hat{y}$, $\hat{p}_y$, $\hat{z}$, and $\hat{p}_z$.

To construct $\hat{L}_z$, we may "quantize" L_z, by simply replacing the classical quantities in L_z with the corresponding quantum-mechanical operators:

$$L_z \longrightarrow \hat{L}_z = \hat{x}\hat{p}_y - \hat{y}\hat{p}_x = i\hbar\left(y\frac{\partial}{\partial x} - x\frac{\partial}{\partial y}\right). \tag{8.3}$$

(Because $\vec{r}$ is the position vector, r_x and r_y are replaced with $\hat{x}$ and $\hat{y}$, respectively.) $\hat{L}_x$ and $\hat{L}_y$ are similarly obtained.

Once we have the explicit form of the angular momentum operators, we can calculate their commutators directly. As discussed in Chapter 7, the commutator of two operators is, in general, another operator, not a constant. For $\hat{L}_x, \hat{L}_y$, and $\hat{L}_z$, we obtain

$$[\hat{L}_x, \hat{L}_y] = i\hbar\hat{L}_z, \qquad [\hat{L}_z, \hat{L}_x] = i\hbar\hat{L}_y, \qquad [\hat{L}_y, \hat{L}_z] = i\hbar\hat{L}_x \ . \tag{8.4}$$

Like any other vector, $\vec{L}$'s components together contain all information about $\vec{L}$ itself. In particular, the magnitude (squared) of $\vec{L}$ is obtained from its components by: $\vec{L}^2 = L^2 = L_x^2 + L_y^2 + L_z^2$. (Here I've used the common notation $\vec{L}^2 = \vec{L} \cdot \vec{L}$.)

The analogous quantum quantity, $\hat{L}^2$, is defined as

$$\hat{L}^2 = \hat{L}_x^2 + \hat{L}_y^2 + \hat{L}_z^2. \tag{8.5}$$

Using Eqs. (8.4) and (8.5) to calculate the commutator of $\hat{L}^2$ with $\hat{L}_x, \hat{L}_y$, and $\hat{L}_z$, leads to

$$[\hat{L}^2, \hat{L}_j] = 0, \qquad j = 1, 2, 3. \tag{8.6}$$

Here x corresponds to 1, y to 2, and z to 3.

These results provide an object lesson in the significance of commutation relations. All of the operators, the three $\hat{L}_j$s along with $\hat{L}^2$, are Hermitian. Thus, the eigenstates of each forms a complete set of states, that is, *any* angular momentum state may be represented as an expansion of the eigenstates of either $\hat{L}^2$, or of any one of the L_js. However, because the commutators in Eq. (8.4) do *not* vanish, any two (different) $\hat{L}_j$s do *not* share a complete set of eigenstates.

All this leads to the following conceptual picture. Suppose the system is in an eigenstate of one of the $\hat{L}_j$s, say $\hat{L}_3$ (i.e. $\hat{L}_z$). This state may also be an eigenstate of $\hat{L}^2$. It *cannot*, however, also be an eigenstate of either $\hat{L}_1$ or $\hat{L}_2$.

Further insight may be gained from the uncertainty relations, which, given Eqs. (8.4) and (8.6), we can simply write down:

$$\Delta L^2 \Delta L_j \geq \frac{|\langle i[\hat{L}^2, \hat{L}_j]\rangle|}{2} = 0, \quad \Delta L_x \Delta L_y \geq \frac{|\langle i[\hat{L}_x, \hat{L}_y]\rangle|}{2} = \frac{|\hbar\langle \hat{L}_z\rangle|}{2}. \quad (8.7)$$

($\Delta L_y \Delta L_z$ and $\Delta L_x \Delta L_z$ are analogous to the $\Delta L_x \Delta L_y$ case.) These relations are not just abstract mathematics: they tell us something important about the physics of quantum-mechanical angular momentum.

The first uncertainty relation in Eq. (8.7) sets a lower bound of zero on $\Delta L^2 \Delta L_j$, and this result is independent of the quantum state. Like all uncertainty relations, this is an *inequality*, so it *doesn't* say that $\Delta L^2 \Delta L_j = 0$, or, therefore, that $\Delta L^2 = 0$ or $\Delta L = 0$; but it does tell us that these equations aren't prohibited by the uncertainty relations (see Section 7.2.2).

Let's now consider $\Delta L_x \Delta L_y$. From the second uncertainty relation in Eq. (8.7), we see that the lower bound set on $\Delta L_x \Delta L_y$ is determined by $\langle L_z \rangle$, an expectation value, which clearly *does* depend on the state. If the system state is such that $\langle L_z \rangle = 0$, then the uncertainty relation alone cannot tell us whether $\Delta L_x \Delta L_y$ is zero or non-zero. But if the system state is such that $\langle L_z \rangle \neq 0$, then $\Delta L_x \Delta L_y > 0$, so both ΔL_x and ΔL_y must be non-zero.

Now suppose, for a particular state, we actually calculate $\Delta L_x \Delta L_y$ directly from the probability distributions (i.e. we do *not* just use the uncertainty relation), and find that $\Delta L_x \Delta L_y = 0$. It should be clear that the *only* way this can happen is if the state is an eigenstate of $\hat{L}_x$ or $\hat{L}_y$, or both. Conversely, if we directly calculate $\Delta L_x \Delta L_y$ for some other state and find that $\Delta L_x \Delta L_y \neq 0$, then that state cannot be an eigenstate of either $\hat{L}_x$ or $\hat{L}_y$.

8.2.2 Eigenstates and Eigenvalues

We know from the postulates of quantum mechanics that if we make a measurement of some observable quantity, we must obtain one of the eigenvalues of the corresponding Hermitian operator. The three operators $\hat{L}_j$ all correspond to observable quantities: each corresponds to angular momentum in some direction, in analogy with the three classical quantities L_x, L_y, and L_z (cf. Eq. (8.2)). In fact, the orientation of our coordinate system is arbitrary, so once we have the eigenvalues and eigenstates for, say, $\hat{L}_z$, we have them for all three $\hat{L}_j$s. The operator $\hat{L}^2$ also corresponds to an observable: the squared magnitude of the total angular momentum, corresponding to the classical quantity L^2.

For our purposes there's little to be gained by working through the eigenvalue problems for $\hat{L}^2$ and $\hat{L}_z$ in detail (and such solutions are readily available elsewhere). What results is the following:

$$\hat{L}^2|l,m\rangle = l(l+1)\hbar^2|l,m\rangle \qquad l = 0,\ \frac{1}{2},\ 1,\ \frac{3}{2},\ 2,\ \frac{5}{2},\ 3,\dots \qquad (8.8)$$

$$\hat{L}_z|l,m\rangle = m\hbar|l,m\rangle \qquad m = -l,\ -l+1,\dots,l-1,\ l \qquad (8.9)$$

Although we've seen eigenvalue equations before, we need to unpack these a bit.

Recall that *two* operators describe angular momentum in quantum mechanics: $\hat{L}^2$ and $\hat{L}_z$, corresponding to the classical L^2 and L_z, respectively. In the ket $|l,m\rangle$, the label l is an index, a bookkeeping tool, that denotes a particular eigenstate of $\hat{L}^2$. While l itself is not the eigenvalue of $\hat{L}^2$ for $|l,m\rangle$, it's very simply related to that eigenvalue (i.e. $l(l+1)\hbar^2$). Similarly, m is an index that denotes a particular eigenstate of $\hat{L}_z$, with corresponding eigenvalue $m\hbar$. Note that the value of l determines the possible values of m. Suppose, for example, that $l = \frac{3}{2}$. Then m could take on the values $-\frac{3}{2}$, $-\frac{1}{2}$, $\frac{1}{2}$, or $\frac{3}{2}$. From Eq. (8.9) we see that for some particular l value there are $2l + 1$ possible m values.

This also suggests why the $|l,m\rangle$ kets have two labels. The $|l,m\rangle$s are eigenstates of both $\hat{L}^2$ and $\hat{L}_z$, but because l doesn't uniquely determine m, and vice versa, both must be given to specify a particular angular momentum eigenstate.

8.2.3 Raising and Lowering Operators

Two other important operators are the *raising and lowering operators*, also called *ladder operators*. The raising operator, $\hat{L}_+$, and the lowering operator, $\hat{L}_-$, are defined as:

$$\hat{L}_+ \equiv \hat{L}_x + i\hat{L}_y, \qquad \hat{L}_- \equiv \hat{L}_x - i\hat{L}_y. \qquad (8.10)$$

When dealing with quantum-mechanical angular momentum, it's conventional to work in the z basis, that is, the basis formed by the eigenstates of $\hat{L}_z$. For example, we take the state $|l,m\rangle$ to be an eigenstate of $\hat{L}_z$, not of $\hat{L}_x$ or $\hat{L}_y$ (unless told otherwise). That's why an eigenvalue equation for $\hat{L}_z$, but not for $\hat{L}_x$ or $\hat{L}_y$, appears in Eq. (8.9). And that's why we have only two, not six, raising and lowering operators: it's implicit that we're working in the z basis.

Now, any angular momentum state can be expanded in the eigenstates of $\hat{L}_z$. Thus, to determine the action of $\hat{L}_+$ and $\hat{L}_-$ on any angular momentum state, we need only determine their action on the eigenstates of $\hat{L}_z$. To do so, first use Eq. (8.4) to show that $[\hat{L}_z, \hat{L}_\pm] = \pm\hbar\hat{L}_\pm$. Then let this operator equation act on an eigenstate of $\hat{L}_z$, and use Eq. (8.9):

$$\left(\hat{L}_z\hat{L}_\pm - \hat{L}_\pm\hat{L}_z\right)|l,m\rangle = \pm\hbar\hat{L}_\pm|l,m\rangle$$

$$\left(\hat{L}_z - m\hbar\right)\hat{L}_\pm|l,m\rangle = \pm\hbar\hat{L}_\pm|l,m\rangle$$

$$\hat{L}_z\left\{\hat{L}_\pm|l,m\rangle\right\} = (m\pm1)\hbar\left\{\hat{L}_\pm|l,m\rangle\right\} \tag{8.11}$$

The curly brackets in the last line emphasize that $\hat{L}_\pm$ acting on the state $|l,m\rangle$ creates a new state, $\hat{L}_\pm|l,m\rangle$, which is an eigenstate of $\hat{L}_z$ with eigenvalue $(m\pm1)\hbar$. Moreover, using Eqs. (8.6) and (8.10), we see that $\hat{L}^2\hat{L}_\pm|l,m\rangle = \hat{L}_\pm\hat{L}^2|l,m\rangle = l(l+1)\hbar^2|l,m\rangle$. That is, $\hat{L}_\pm|l,m\rangle$ is an eigenstate of $\hat{L}^2$ with eigenvalue $l(l+1)\hbar^2$. Thus, up to a multiplicative constant, the new state must be $|l,m\pm1\rangle$ (see Eq. (8.9)).

The monikers "raising operator," for $\hat{L}_+$, and "lowering operator," for $\hat{L}_-$, now become clear: the two operators, acting on an eigenstate of $\hat{L}_z$, raise or lower, respectively, the value of m by 1. Although we won't work it out, the multiplicative constant is not unity. Rather, we have:

$$\hat{L}_\pm|l,m\rangle = \sqrt{l(l+1) - m(m\pm1)}\ \hbar\ |l,m\pm1\rangle. \tag{8.12}$$

Note that this multiplicative constant guarantees that m cannot be raised above, or lowered below, the limit given by Eq. (8.9), that is, $|m| \le l$.

It's worthwhile introducing the angular momentum raising and lowering operators—even though we won't need them—in part because similar operators appear elsewhere in quantum mechanics. In the harmonic oscillator, operators analogous to $\hat{L}_+$ and $\hat{L}_-$ raise and lower the oscillator's energy (not angular momentum) eigenstate. In quantum electrodynamics—the quantum field theory of electromagnetism—the electromagnetic field consists of a collection of photons. Mathematically, this field may be represented by a collection of harmonic oscillators. The oscillator's raising and lowering operators are now called *creation and annihilation operators*, respectively, because they either create or annihilate quanta of the electromagnetic field (photons).[7]

8.3 Physical Interpretation

8.3.1 Measurements

In quantum mechanics, one can often attain a better understanding by considering what actual measurements would yield. Considering measurement results for L^2 and L_z will help illuminate quantum angular

---

[7] For more on the harmonic oscillator and related topics, see Appendix C.

momentum.[8] Moreover, we'll make contact again with the postulates of quantum mechanics, from Chapter 2, and the statistical interpretation, from Chapter 3.

In many texts, discussions of angular momentum measurements center on the *Stern–Gerlach* (SG) machine—essentially a non-uniform magnetic field. A system with electric charge and quantum angular momentum—perhaps a silver atom, as in Stern and Gerlach's original 1922 experiment—can possess a magnetic moment that interacts with an external magnetic field. If such an atom passes through an SG machine, the atom's magnetic moment interacts with the magnetic field such that the atom (more correctly, the atom's quantum state) is deflected in different directions depending on the value of the angular momentum. The atom then impacts a screen.[9]

For present purposes, the details of SG experiments, or of any other means to perform angular momentum measurements, are of little interest. It's enough to say that we have a device, a "black box", that somehow performs measurements of quantum angular momentum along whatever axis we choose.

The simplest non-trivial case is that of $l = 1/2$, and thus either $m = -1/2$ or $m = 1/2$. Suppose the system is prepared with $m = 1/2$, so our state, in the L_z basis, is $\left|\frac{1}{2}, \frac{1}{2}\right\rangle$. From Eqs. (8.8) and (8.9) we see that

$$\hat{L}_z \left|\tfrac{1}{2}, \tfrac{1}{2}\right\rangle = \tfrac{\hbar}{2} \left|\tfrac{1}{2}, \tfrac{1}{2}\right\rangle, \qquad \hat{L}^2 \left|\tfrac{1}{2}, \tfrac{1}{2}\right\rangle = \tfrac{3\hbar^2}{4} \left|\tfrac{1}{2}, \tfrac{1}{2}\right\rangle. \tag{8.13}$$

So, for the state $\left|\frac{1}{2}, \frac{1}{2}\right\rangle$, we will definitely obtain $\hbar/2$ if we measure L_z, and $3\hbar^2/4$ if we measure L^2.

What if we adjusted our black box to measure L_x rather than L_z? $\hat{L}_x$ is a Hermitian operator, like $\hat{L}_z$, so its eigenstates must form a complete set. For this $l = 1/2$ case there are only two eigenstates. These form the complete set and are exactly analogous to the eigenstates of $\hat{L}_z$, so we denote them $\left|\frac{1}{2}, \frac{1}{2}\right\rangle_x$ and $\left|\frac{1}{2}, -\frac{1}{2}\right\rangle_x$. We can write our system state, $\left|\frac{1}{2}, \frac{1}{2}\right\rangle$, in terms of this complete set: $\left|\frac{1}{2}, \frac{1}{2}\right\rangle = c_+ \left|\frac{1}{2}, \frac{1}{2}\right\rangle_x + c_- \left|\frac{1}{2}, -\frac{1}{2}\right\rangle_x$. In addition, a pair of eigenvalue equations just like those in Eq. (8.13) hold.

So, what happens when we measure L_x?[10] We can only say that we'll obtain $\hbar/2$, the eigenvalue of $\hat{L}_x$ for $\left|\frac{1}{2}, \frac{1}{2}\right\rangle_x$, with probability $|c_+|^2$, and $-\hbar/2$, the eigenvalue of $\hat{L}_x$ for $\left|\frac{1}{2}, -\frac{1}{2}\right\rangle_x$, with probability $|c_-|^2$. If the measured result is $\hbar/2$, the system is left in the corresponding eigenstate,

---

[8] Measurable quantities, such as L^2 and L_z, *correspond* to operators, but aren't operators themselves; thus, I omit "hats" from them. For the same reason, uncertainties in measurable quantities, such as ΔL^2 and ΔL_j in Eq. (8.7), are *not* written as operators.

[9] In-depth discussions of SG machines/experiments are readily found. See, for example, Bohm (1951), Feynman (1965), or Townsend (1992).

[10] Very similar arguments apply for L_y.

$\left|\frac{1}{2},\frac{1}{2}\right\rangle_x$; if the result is $-\hbar/2$, the system is left in $\left|\frac{1}{2},-\frac{1}{2}\right\rangle_x$. (See Postulate 3 of Chapter 2.)

Now consider measuring L^2. The fact that $\hat{L}^2 = \hat{L}_x^2 + \hat{L}_y^2 + \hat{L}_z^2$ suggests measuring, squaring, and adding L_x, L_y, and L_z to obtain a measured value for L^2. Quantum-mechanical measurements are always made on a system in some particular quantum state. For the case at hand, "measuring L^2" really means "measuring L^2 in the state $\left|\frac{1}{2},\frac{1}{2}\right\rangle$." So let's measure L_x, L_y, and L_z on our system state, $\left|\frac{1}{2},\frac{1}{2}\right\rangle$.

But there's a problem. If we measure, say, L_x first, the system is left in either $\left|\frac{1}{2},\frac{1}{2}\right\rangle_x$ or $\left|\frac{1}{2},-\frac{1}{2}\right\rangle_x$. So the next measurement is *not* being made on the original state, $\left|\frac{1}{2},\frac{1}{2}\right\rangle$. If we measure L_z first, the state does not change, because $\left|\frac{1}{2},\frac{1}{2}\right\rangle$ is an eigenstate of $\hat{L}_z$. However, L_x and L_y must still be measured, and the system will *not* be in $\left|\frac{1}{2},\frac{1}{2}\right\rangle$ for the final measurement. And in general, an initial state need not be an eigenstate of $\hat{L}_x, \hat{L}_y$, or $\hat{L}_z$.

The solution is to re-prepare the system in the state $\left|\frac{1}{2},\frac{1}{2}\right\rangle$ after each measurement, so that all measurements are performed on that state. Yet for $l = 1/2$, none of this really matters. For *any* $l = 1/2$ state, $\hbar/2$ and $-\hbar/2$ are the only eigenvalues, and thus the only possible measured values, for L_x, L_y, and L_z. For *every* measurement of L_x, L_y, or L_z, then, in *any* $l = 1/2$ state, we get $L_x^2 = L_y^2 = L_z^2 = \hbar^2/4$; thus $L^2 = L_x^2 + L_y^2 + L_z^2 = 3\hbar^2/4$.

But consider a case where $l > 1/2$, say $l = 1$. For each operator, $\hat{L}_x, \hat{L}_y$, and $\hat{L}_z$, there are now three eigenstates. And unlike the $l = 1/2$ case, the squares of $\hbar, -\hbar$, and 0—the various eigenvalues—are *not* all equal.

Now prepare an $l = 1$ state and measure L_x, re-prepare the same state and measure L_y, and then do the same for L_z. If the state is a superposition in the L_x, L_y, and L_z bases, each L_x, L_y, or L_z measurement can yield $\hbar, -\hbar$, or 0. Thus, L_x^2, L_y^2, and L_z^2 can be $\hbar^2$ or 0, but which we obtain in a particular measurement is determined only probabilistically, and is independent of previous measurements. Calculating $L_x^2 + L_y^2 + L_z^2$ from such a sequence of L_x, L_y, and L_z measurements may or may not yield $2\hbar^2$, the eigenvalue of $\hat{L}^2$. In general, we'll obtain $2\hbar^2$ only by averaging over many sets of L_x, L_y, and L_z measurements. For $l > 1/2$, it seems we can't make a single measurement of L^2.

In fact, measurements of L^2 are routinely performed—but by somewhat indirect methods. Chemists and physicists measure energy levels of atoms, molecules, and nuclei. These energies can be expressed in terms of the quantum number l, so a measurement of energy effectively constitutes a measurement of l, and thus of L^2. My goal here has not been to argue that L^2 is unmeasurable, but to use measurement results to better understand quantum angular momentum—and quantum mechanics generally.

It's worth noting that the discussion of measurements in this section is consistent with, and perhaps best understood in terms of, the statistical interpretation. That is, we treat the quantum state as determining the probability for obtaining a particular value upon measuring some observable—nothing more, and nothing less.

8.3.2 Relating L^2 and L_z

Inspection of Eqs. (8.8) and (8.9) reveals something odd. We interpret $\hat{L}^2$ as the operator corresponding to the square of the total angular momentum, while $\hat{L}_z$ corresponds to the projection of the angular momentum onto the z axis. Yet even if we choose the maximum m value for a given l value, L^2 is greater than L_z^2. That is, the magnitude of the total angular momentum is greater than the projection of the angular momentum onto the z axis.

In terms of measurements, it's simplest to see how this arises for the $l = 1/2$ case. From Section 8.3.1, measurement of L_x, L_y, or L_z in *any* $l = 1/2$ state yields $L_x^2 = L_y^2 = L_z^2 = \hbar^2/4$, so $L^2 = L_x^2 + L_y^2 + L_z^2 = 3\hbar^2/4$. Evidently the magnitude of the total angular momentum is $\sqrt{3}\hbar/2$, but the maximum projection is only $\hbar/2$. How can we make sense of this?

One approach is to try to think of L^2 as the squared magnitude of an ordinary vector, and of L_z as the projection of that vector onto the z axis. Then the argument, roughly, is that there must be projections onto the x and y axes as well, since a measurement of L_x or L_y cannot yield zero. For an ordinary vector, such projections can exist only if the vector does not lie along the z axis—and this, in turn, implies that $L_z^2 < L^2$.[11]

But there's a price to be paid in this approach: quantum angular momentum *isn't* an ordinary vector, so one can't attain a truly correct understanding through this vector model, and misconceptions may be introduced. Let's try another approach to understanding the fact that $L_z^2 < L^2$.

Given the classical quantity $L^2 = L_x^2 + L_y^2 + L_z^2$, we're certainly free to *define* the quantum operator $\hat{L}^2 \equiv \hat{L}_x^2 + \hat{L}_y^2 + \hat{L}_z^2$. But that doesn't necessarily justify attaching the same *meaning* to the classical L^2 and the quantum $\hat{L}^2$. Consider the physically meaningful quantities, in the sense of physically *measurable* quantities, in the classical and quantum cases.

Classically, angular momentum—like position or energy—is a basic quantity and enjoys a well-defined existence. But it's really incorrect to refer to the position, energy, or angular momentum of a quantum system—even though doing so often causes little trouble.[12] In quantum mechanics, it's the *state* that's well defined—and the state yields only *probabilities* for observables such as angular momentum. Quantum and classical angular momentum, though related, simply aren't the same—quantum angular momentum isn't an ordinary vector, and we can't assume it will behave as one.

---

[11] The uncertainty relations are often offered as further justification.
[12] It *is* legitimate if the system is in an eigenstate of the relevant quantity.

But then how *does* it behave? If the classical vector $\vec{L}$ points along the z axis, then the x and y components of $\vec{L}$ vanish identically. The analogous quantum case is that of an eigenstate of $\hat{L}_z$, such as $|\frac{1}{2}, \frac{1}{2}\rangle$. If we expand $|\frac{1}{2}, \frac{1}{2}\rangle$ in the eigenstates of L_x, we get:[13]

$$\left|\tfrac{1}{2}, \tfrac{1}{2}\right\rangle = \tfrac{1}{\sqrt{2}} \left|\tfrac{1}{2}, \tfrac{1}{2}\right\rangle_x + \tfrac{1}{\sqrt{2}} \left|\tfrac{1}{2}, -\tfrac{1}{2}\right\rangle_x. \tag{8.14}$$

Quite generally, if we expand a quantum state in another basis, the basis states (i.e. their expansion coefficients) *can't* all vanish; if they did, the state itself would vanish in the new basis! The coefficients of $|\frac{1}{2}, \frac{1}{2}\rangle_x$ and $|\frac{1}{2}, -\frac{1}{2}\rangle_x$—the x "components" of the state—do not vanish, because a quantum state is not an ordinary vector in real space.

However, a measurement of L_x must yield $\hbar/2$ or $-\hbar/2$, so $L_x^2 > 0$, which implies that $L_z^2 < L^2$. Note also, from Eq. (8.5), that a measured value of L^2 satisfies: $L^2 = L_x^2 + L_y^2 + L_z^2$.

A geometrical approach is helpful. The vector $\vec{G}$ in Figure 8.2(a) clearly has non-zero projections onto the x and y axes. Rotating $\vec{G}$ to align with the z axis yields $\vec{G}'$, whose projections onto the x and y axes are clearly zero. This is precisely how a classical angular momentum vector would behave.

Now, for each operator, $\hat{L}_x, \hat{L}_y,$ and $\hat{L}_z$, there exists a complete set of basis states. For the simple $l = 1/2$ case, two orthonormal basis states exist for each operator—that is, for any *one* axis in real space, there are *two* basis states, and thus *two* directions in Hilbert space. Figure 8.2(b) illustrates this heuristically. *Any* $l = 1/2$ state, even an eigenstate of $\hat{L}_z$, could be written strictly in terms of the basis states of, say, $\hat{L}_x$.

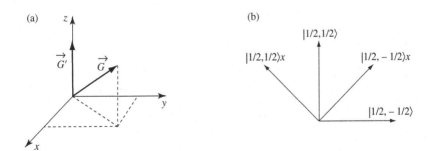

Fig. 8.2 (a) The vector $\vec{G}$, and its rotated counterpart $\vec{G}'$, in real space; (b) heuristic representation of basis states of $\hat{L}_x$ and $\hat{L}_z$ in Hilbert space for $l = 1/2$.

---

[13] That is, the coefficients c_+ and c_- of Section 8.3.1 actually satisfy $c_+ = c_- = 1/\sqrt{2}$.

Thus, the argument that $L_z^2 < L^2$ because there must be a projection onto the x and y axes is flawed because the situation in real space differs from that in Hilbert space. Real space is three-dimensional, and we clearly cannot represent an arbitrary vector in real space by considering only one axis. But for quantum angular momentum, any one direction in real space has a complete set of states, $2l + 1$ in number, associated with it. For $l > 1/2$, things aren't quite so simple as in the discussion above, but the additional complication is more a matter of degree than of principle, and the basic concepts as developed for the $l = 1/2$ case remain valid.

8.4 Orbital and Spin Angular Momentum

8.4.1 Orbital Angular Momentum

Classically, it's easy to imagine an object with angular momentum due both to its orbit and due to its spinning. The Earth, for example, orbits the Sun, but also spins on its axis. Both motions contribute to the angular momentum, and both are defined by $\vec{L} = \vec{r} \times \vec{p}$.

Angular momentum due to an orbit and due to an actual spinning object can occur in quantum mechanics, also. But in quantum mechanics, these would both be treated as orbital angular momentum—spinning is, after all, just an object orbiting on its own axis. In quantum mechanics, *spin angular momentum*, or just *spin*, does *not* refer to an actually spinning object. The two types of quantum angular momentum, orbital and spin, rest on profoundly different conceptual bases.

We didn't (and won't) solve the angular-momentum eigenvalue problem in detail. Nevertheless, an outline of how it can be solved illuminates the difference between orbital and spin angular momentum in quantum mechanics.

One approach to this eigenvalue problem is through the position-space representations of the operators $\hat{L}_z$ (see Eq. (8.3)) and $\hat{L}^2$. Because such operators appear in terms of spatial derivatives, they clearly must act on a quantum state that is itself written in terms of spatial variables.[14]

If, using these operators, we set up and solve the eigenvalue equations of Eqs. (8.8) and (8.9) (which become differential equations), we obtain the eigenstates (in position representation) and eigenvalues of $\hat{L}_z$ and $\hat{L}^2$. However, the half-integer values of l, and thus also of m, that appear in Eqs. (8.8) and (8.9) *are disallowed*. Why?

Because we are working in position-space, the solutions must be continuous functions of position. This implies the imposition of boundary conditions that force our solutions to be physically acceptable. But these conditions can be satisfied only by disallowing the half-integer values

---

[14] In practice, one uses the spherical coordinates r, θ, and ϕ, rather than x, y, and z.

of l. Restated, insisting that the position-space solutions be physically reasonable requires us to discard the half-integer values of l.

Note that in the approach to the angular-momentum eigenvalue problem just outlined, we are working in position space—the concept of the particle's position probability forms the essential underpinning. This, we will see, is in striking contrast to the situation if one adopts the following alternative approach to the problem.

8.4.2 Spin Angular Momentum

Quantum mechanics is replete with mysteries, and spin angular momentum (or just "spin") is one of the deepest. A broad-brushstroke discussion of an alternative way in which the angular-momentum eigenvalue problem may be solved sheds light on why spin is so abstract and mysterious.

This alternative approach doesn't rely on the Schrödinger equation or position representation. Recall that in Section 8.2.1 we obtained the angular-momentum operators, and then the commutators, from the operators $\hat{x}, \hat{p}_x$, etc.—but I said that the commutators may also be obtained through an approach based on continuous transformations (in particular, by studying the behavior of quantum states under rotations). Moreover, once they are so obtained, the eigenvalue problem can be solved directly from the commutation relations (how is not our concern). Remarkably, the half-integer values of l now *are* allowed—but what can they mean?

The eigenvalues corresponding to half-integer l values characterize *spin* angular momentum. But how can it be that spin apparently violates our requirements for the position representation of angular momentum eigenstates? Because, unlike orbital angular momentum, quantum-mechanical spin angular momentum *doesn't have a position representation*; it defies physical pictures. Yet the eigenvalues, commutators, raising and lowering operators, etc., for spin behave just like those for "ordinary", orbital angular momentum. Spin is therefore often called *intrinsic* angular momentum—it carries all the hallmarks of a quantum angular momentum, but denies us the physical picturability we might hope for. Subtle is the Lord.

8.5 Review

Our study of angular momentum has taken us to new heights (or is it depths?) of abstraction. A brief review might help make sense of it all.

Starting from classical angular momentum, we saw how the quantum operators $\hat{L}_x, \hat{L}_y, \hat{L}_z$, and $\hat{L}^2$ are constructed from the position and linear momentum operators. These led to the angular momentum commutators and uncertainty relations, from which we concluded that the various $\hat{L}_j$s do *not* share eigenstates, but each $\hat{L}_j$ *does* share eigenstates with $\hat{L}^2$.

The eigenvalue equations for $\hat{L}^2$ and $\hat{L}_z$, and the corresponding quantum numbers l and m, were then introduced. I also introduced the raising and lowering operators, which either increase or decrease the value of the quantum number m by 1.

A discussion of angular momentum measurements followed—not to delve into the details of measuring devices, but to illustrate schematically how such measurements could be carried out, and to attach physical meaning to quantum angular momentum. Our discussion also set much of what we've learned about bases, superpositions, probabilities, and the statistical interpretation into operational context.

Finally, I outlined the deep differences between orbital and spin angular momentum. Treating angular momentum in terms of the operators $\hat{x}, \hat{p}_x$, etc. is useful and illustrative, but not general. An alternative approach, using group theory and commutation relations, includes half-integer values of l and therefore naturally accounts for both orbital and spin angular momentum. Spin is an abstract angular momentum which has no position-space representation, and which cannot be associated with the actual spinning of an object.

Quantum-mechanical angular momentum is immensely important: it's crucial in atomic, molecular, nuclear, and particle physics. The coupling of two or more systems to form states of total angular momentum plays a key role in many applications, and provides a prime example of one of the more fascinating topics in physics: quantum entanglement and non-locality. Although we've only scratched the surface of quantum-mechanical angular momentum, the fundamentals are now on the table.

Practical utility aside, however, quantum-mechanical angular momentum forces us to directly confront the abstract nature of the quantum world. Einstein said that the Lord is subtle but not malicious. Yet, some years later, perhaps while immersed in a particularly deep and abstract problem, he returned to the subject. "I have second thoughts," he remarked to his colleague Valentine Bargmann, "Maybe God *is* malicious."

8.6   Problems

1. Obtain the angular momentum commutation relations from the operators by direct calculation.

2. Suppose the operators $\hat{A}$, $\hat{B}$, and $\hat{C}$ correspond to observables A, B, and C, respectively, and that they satisfy the following commutation relations:

$$[\hat{A}, \hat{B}] = 0, \qquad [\hat{A}, \hat{C}] = 0, \qquad [\hat{B}, \hat{C}] \neq 0.$$

What do these commutation relations tell us about the eigenstates of the various operators? In particular, is it possible for any state to be an eigenstate of more than one of these operators? Can any state be an eigenstate of all three operators? Discuss what all this means for various sequential measurements of the observables corresponding to $\hat{A}$, $\hat{B}$, and $\hat{C}$?

3. This problem pertains to a spin $\frac{1}{2}$ system.

(i) Construct spin$\frac{1}{2}$ matrix representations of $\hat{S}_+$ and $\hat{S}_-$; then use those matrices to construct the matrix operator $\hat{S}_+\hat{S}_-$.

(ii) Using your matrix for $\hat{S}_+\hat{S}_-$, operate on the following spin$\frac{1}{2}$ states:

(a) $|\frac{1}{2}, \frac{1}{2}\rangle$

(b) $|\frac{1}{2}, -\frac{1}{2}\rangle$

(c) $|\chi\rangle = \frac{1}{\sqrt{3}}|\frac{1}{2}, \frac{1}{2}\rangle + \sqrt{\frac{2}{3}}|\frac{1}{2}, -\frac{1}{2}\rangle$

(iii) Explain why your results make sense, that is, why they are what you expect, for each of the three cases.

4. This problem pertains to a spin 1 system.

(i) Write the angular momentum eigenstates (in the conventional z basis).

(ii) Construct an operator $\hat{I}$ that transforms any state into the z basis.

(iii) Show that your operator does what it's supposed to do by operating on an arbitrary spin 1 state, denoted $|\Psi\rangle$, with $\hat{I}$. Explain why your result is what is to be expected, that is, how can you see that it's correct.

5. This problem pertains to a spin $\frac{1}{2}$ system.

(i) Obtain the uncertainty relation for S_x and S_y, if the system state is: $|\Phi\rangle = \frac{1}{\sqrt{3}}\left(|+z\rangle - i\sqrt{2}|-z\rangle\right)$.

(ii) How would you expect the uncertainty relation between S_x and S_y, for the state of Part i, to be manifested physically? That is, what measurement(s) would you make in the lab to check this relation? What would you find, experimentally? Be concise but clear.

(iii) For the state $|+z\rangle$, we have $\Delta S_x \Delta S_z = 0$. By considering the nature of $|+z\rangle$ in relation to the observable(s) in question, explain why this must be true. A few carefully worded sentences should be sufficient.

6. You have available a beam of spin $\frac{1}{2}$ particles moving in the $+y$ direction, in random spin states. Starting with this beam, carefully describe an experiment that could be used to determine ΔS_z for the state $|+n\rangle$, where n denotes some arbitrary direction in the xz plane. (i.e., describe an experiment the results of which could be used to construct ΔS_z for the state.) Include a step-by-step description of how the measurements are to be carried out.

7. The operators $\hat{J}_+$ and $\hat{J}_-$ are given by $\hat{J}_\pm = \hat{J}_x \pm i\hat{J}_y$.

(i) Construct spin-1 matrix representations of $\hat{J}_+$ and $\hat{J}_-$. Then operate on the spin-1 states $|1\rangle$ and $|-1\rangle$ with $\hat{J}_-\hat{J}_+$, and with $\hat{J}_+\hat{J}_-$; explain why your results make physical sense.

(ii) Suppose you were to operate on the state $|\psi\rangle = \frac{1}{\sqrt{2}}|1\rangle + \frac{1}{\sqrt{2}}|0\rangle$ with each operator, $\hat{J}_-\hat{J}_+$, and $\hat{J}_+\hat{J}_-$. Think about what you *should* get in each case, and write down your prediction. Then carry out the operations. Explain why your results make sense.

(iii) Are any of the states that you acted on in Parts (i) and (ii) eigenstates of either $\hat{J}_-\hat{J}_+$ or $\hat{J}_+\hat{J}_-$? How do you know?

8. For a spin 1 system, the raising and lowering operators (in the z basis) are

$$\hat{S}_+ = \sqrt{2}\hbar \begin{pmatrix} 0 & 1 & 0 \\ 0 & 0 & 1 \\ 0 & 0 & 0 \end{pmatrix}, \qquad \hat{S}_- = \sqrt{2}\hbar \begin{pmatrix} 0 & 0 & 0 \\ 1 & 0 & 0 \\ 0 & 1 & 0 \end{pmatrix}$$

(i) Using $\hat{S}_+$, $\hat{S}_-$, and Eq. (8.10), obtain the matrix representations of $\hat{S}_x$ and $\hat{S}_y$, in the z basis, for a spin 1 particle.

(ii) From the commutation relations and the results of Part (i), find the matrix $\hat{S}_z$ in the z basis.

9
The Time-Independent Schrödinger Equation

They do not forgive Epicurus for having supposed, to account for the most impor-
tant things, an event as small and insignificant as the minute declination of a
single atom, in order to introduce with much cunning the celestial bodies, the liv-
ing beings, and destiny, and for the purpose that our free will not be annihilated.
Plutarch[1]

Although our atomic concepts are nothing like those of Epicurus, our modern understanding of atoms largely *does* explain celestial bodies and, to a lesser extent, living beings. As for explaining destiny and free will—don't get your hopes up.

It's all too easy to take our scientific achievements for granted. Basic atomic structure is now routinely taught in high-school. But our understanding of atoms is, in fact, the crowning achievement in a 2,500 year quest to understand the constituents of the physical world. Yet almost all of that understanding has come about since the 1920s—that is, since the advent of quantum mechanics.

The time-independent Schrödinger equation (TISE) is the quantum tool that ushered in that understanding. Because it explains atomic, molecular, and even nuclear structure, much modern physics and chemistry centers on this equation. Many quantum texts introduce the TISE early on, and greatly emphasize its application. I have not. Why?

Important as it is, the TISE concerns only one area of quantum mechanics. Our overarching goal has not been to grasp quantum mechanics' applications, but its *structure*—the conviction being that this is best accomplished through a general treatment. With that general treatment now in hand, we can understand time-independent systems within the broader context of the quantum formalism.

[1] Pullman (1998), p. 39.

9.1   An Eigenvalue Equation for Energy

Although you may have heard that the Schrödinger equation is the key to quantum mechanics, I've said precious little about it, apart from a brief introduction in Chapter 2. The time-dependent Schrödinger equation is

$$i\hbar \frac{d\Psi}{dt} = \hat{H}\Psi. \tag{9.1}$$

Here Ψ is a quantum state (I could have written it as the ket $|\Psi\rangle$). Equation (9.1) is a general law of quantum mechanics. Thus, no reference is made to the type of quantum system that Eq. (9.1) describes.

As prelude, consider Newton's second law:[2] $\vec{F}_{net} = m\vec{a}$, where $\vec{F}_{net}$ is the net force on an object, m is its mass, and $\vec{a}$ is its acceleration. The second law's simple appearance is deceptive: it can describe an endless variety of physical systems, corresponding to different net forces. And depending on the net force, the second law can easily become difficult—even impossible—to solve exactly.

Similarly, Eq. (9.1) may look simple, and not particularly interesting. But the Hamiltonian operator, $\hat{H}$, can take on an endless variety of forms. So, just what *is* this $\hat{H}$?

In classical physics, the term "analytical mechanics" refers to various formulations of mechanics which do not rely on Newton's second law. In one of these, Hamiltonian mechanics, a central role is played by the Hamiltonian, $H = H(x, y, z, p_x, p_y, p_z, t)$, a function of the coordinates, the momenta, and possibly the time.[3] In-depth discussions of Hamiltonian mechanics are readily available; all we need to know is that, for almost all cases, $H = K + V$. That is, H is simply the sum of the kinetic and potential energy functions, K and V.

To obtain the quantum operator $\hat{H}$, we "quantize" the classical H by substituting in appropriate quantum operators. Then we can write down Eq. (9.1),

$$i\hbar \frac{\partial \Psi(x, y, z, t)}{\partial t} = \hat{H}\Psi(x, y, z, t), \tag{9.2}$$

and substitute in our particular $\hat{H}$. In one sense, Eq. (9.2) is a differential equation; in another, it's a limitless collection of differential equations! (If my use of a partial derivative confuses you here, see note 11 of Chapter 5.)

Let's first consider the simplest possible case: a free particle in one-dimension. The fundamental (and fundamentally important!) classical relation $\vec{F} = -\nabla V$ becomes, in one-dimension, $F = -\frac{dV}{dx}$. For a free

---

[2] The second law actually is: $\vec{F}_{net} = d\vec{p}/dt$. But for constant m, this equation reduces to the familiar form $\vec{F}_{net} = m\vec{a}$, which is adequate for our purposes.

[3] Here H is written in terms of Cartesian coordinates, but much of analytical mechanics' power involves using alternative coordinate systems—a subtle topic that I won't discuss.

particle, $F = 0$, so $\frac{dV}{dx} = 0$, that is, V is a constant. With no loss of generality, we can choose $V = 0$.[4] The classical Hamiltonian is then just $H = K = p^2/2m$. To quantize H, we substitute the operator $\hat{p}$ for p, that is, $p \rightarrow \hat{p} = -i\hbar\frac{\partial}{\partial x}$. Thus, $\hat{H} = \frac{-\hbar^2}{2m}\frac{\partial^2}{\partial x^2}$, and Eq. (9.2) becomes

$$\frac{\partial}{\partial t}\Psi(x,t) = \frac{i\hbar}{2m}\frac{\partial^2}{\partial x^2}\Psi(x,t). \tag{9.3}$$

If the force is non-zero, things get more interesting. For a classical harmonic oscillator, the force is $F = -kx$, with k a constant. Then $-kx = -\frac{dV}{dx}$, or $V = \frac{1}{2}kx^2$, so we have $H = \frac{p^2}{2m} + \frac{1}{2}kx^2$. For the quantum oscillator, $\hat{H} = \frac{-\hbar^2}{2m}\frac{\partial^2}{\partial x^2} + \frac{1}{2}kx^2$, and the Schrödinger equation becomes

$$i\hbar\frac{\partial}{\partial t}\Psi(x,t) = \frac{-\hbar^2}{2m}\frac{\partial^2}{\partial x^2}\Psi(x,t) + \frac{1}{2}kx^2\Psi(x,t). \tag{9.4}$$

In principle, an infinite variety of classical forces, and thus of potential energy functions V, could exist—each corresponding to a different $\hat{H}$. These classical forces, and Vs, could also be time-dependent, leading to time-dependent $\hat{H}$s. But to obtain the TISE, we specify that $\hat{H}$ is independent of t, that is: $\hat{H} \neq \hat{H}(t)$.

From above we see—even for the simplest case of a free particle—that when we substitute in an actual $\hat{H}$ we obtain a *partial* differential equation. But because we now take $\hat{H} \neq \hat{H}(t)$, we can apply *separation of variables* to our differential equation.[5] We assume $\Psi(x,t)$ may be written as the product of a function T, which depends only on t, and another function ψ, which depends only on x: $\Psi(x,t) = T(t)\psi(x)$. Then Eq. (9.1) becomes

$$i\hbar\psi(x)\frac{dT(t)}{dt} = T(t)\hat{H}\psi(x)$$

$$i\hbar\frac{1}{T(t)}\frac{dT(t)}{dt} = \frac{1}{\psi(x)}\hat{H}\psi(x). \tag{9.5}$$

Here I've used the fact that $\frac{d}{dt}$ does not alter $\psi(x)$, and $\hat{H}$ does not alter $T(t)$. Because the left-hand side of the (second) equation above now depends only on t, and the right-hand side depends only on x, both sides must be equal to the same constant. As such, we can write

$$\frac{1}{\psi(x)}\hat{H}\psi(x) = E$$

$$\hat{H}\psi(x) = E\psi(x), \tag{9.6}$$

Equation (9.6) is the TISE.

---

[4] We know from elementary classical physics that changing the zero point of potential energy has no effect on the physics.

[5] For details, consult most any discussion of partial differential equations.

Now, what's this E that appears in Eq. (9.6)? The separation of variables process only tells us that E is some constant, not its physical significance. Clearly, however, Eq. (9.6) is an eigenvalue equation, with E the eigenvalue. We know that the classical H corresponds to the total energy, so we recognize Eq. (9.6) as an eigenvalue equation for energy.

We saw that Newton's second law, $\vec{F}_{net} = m\vec{a}$, is deceptively simple—and so is Eq. (9.6). This is due in part to the fact that Eq. (9.6) is restricted to one dimension in Cartesian coordinates. The key point, however, is that $\hat{H}$ can take on an endless variety of forms, each corresponding to a different differential equation: Eq. (9.6) is very rich indeed!

Equation (9.6) arises from the spatially dependent part of Eq. (9.5); another equation arises from its time-dependent part. Because, again, both sides of Eq. (9.5) must equal the same constant, E, we may write

$$\frac{dT(t)}{dt} = \frac{-i}{\hbar} E T(t). \tag{9.7}$$

This first-order differential equation is easily integrated, yielding the solution

$$T(t) = A \exp(-iEt/\hbar), \tag{9.8}$$

where A is an arbitrary constant. The full solution to the time-dependent Schrödinger equation (but still with $\hat{H} \neq \hat{H}(t)$) is, therefore,

$$\Psi(x,t) = T(t)\psi(x) = A \exp(-iEt/\hbar)\psi(x). \tag{9.9}$$

I'll discuss this expression for $\Psi(x,t)$ later in this chapter. Moreover, it will be central to our discussion of time-dependent quantum systems in Chapter 11.

9.2 Using the Schrödinger Equation

9.2.1 Conditions on Wavefunctions

I remarked that Eq. (9.1) is a general law of quantum mechanics. In Eq. (9.2), however, a restriction has already been imposed: the assumption that we're working in position representation, that is, $\Psi = \Psi(x,y,z,t)$. This occurred because we started with the classical $H = H(x,y,z,t)$.

If we're working in position representation, we're working with wavefunctions. Some books concentrate heavily on wavefunctions; I have not. The reason, again, is that our goal has been to develop an understanding of the general structure of quantum mechanics, and wavefunctions are but one particular type of quantum state (albeit an important type), in the sense that they are written in one particular representation.

When partial differential equations appear in *classical* physics, we generally have to impose boundary conditions—that is, the actual physical constraints that the system satisfies—to obtain physically acceptable solutions. For example, if we wish to determine the electric potential in some region, we solve Laplace's partial differential equation—but to eliminate physically unacceptable solutions, we must impose boundary conditions.

Wavefunctions, also, satisfy physically reasonable conditions. First, we insist that wavefunctions, like other quantum states, be normalized: the total probability must equal 1. We also insist that a wavefunction, along with its first derivative, be continuous.[6]

With these constraints on acceptable wavefunctions in hand, let's briefly outline how the TISE is typically used in practice.

- Determine the classical Hamiltonian, H, for the system of interest.
- Obtain the Hamiltonian operator, $\hat{H}$, corresponding to H.
- Substitute $\hat{H}$ into the time-independent Schrödinger equation (Eq. (9.6)).
- Solve the resulting differential equation.
- Impose continuity of the wavefunction and its first derivative.
- Normalize the wavefunction

It is the imposition of continuity on the solutions and their first derivatives that leads to a discrete, or "quantized", set of eigenvalues and eigenstates.[7]

9.2.2   An Example: the Infinite Potential Well

In Section 4.4.2 we briefly discussed the infinite potential well, or "particle-in-a-box." This system consists of a potential that's zero for $0 < x < L$, and infinite elsewhere. Because solving the Schrödinger equation typically becomes very challenging very quickly, I will focus on this system, since it exhibits the generic features I want to illustrate while remaining mathematically simple.

The infinite potential well clearly is *not* a physically realistic system: real potentials cannot be infinite, nor can they change infinitely fast. Because of these unphysical features, the wavefunction conditions for this system differ from those discussed above.

- Continuity of the wavefunction still applies, but continuity of its first derivative does not (see note 6).

---

[6] Write Eq. (9.6) as: $d^2\psi(x)/dx^2 = 2m/\hbar^2(V - E)\psi(x)$. If V, E, and $\psi(x)$ are finite, $d^2\psi(x)/dx^2$ must be finite; this holds only if $d\psi(x)/dx$ is continuous. See Bohm (1951), p. 232. For infinite V (as in Section 9.2.2), this argument fails, so $d\psi(x)/dx$ can be discontinuous.

[7] "Quantize" is one of those slightly dangerous words in physics whose meaning depends on context. Here it means that the energy eigenvalues constitute a discrete set, rather than a continuous distribution. Earlier, however, "quantize" meant creating a quantum-mechanical operator from the corresponding classical quantity (such as the Hamiltonian).

- The wavefunction vanishes identically where the potential is infinite.[8]

With those modifications in mind, let's plunge ahead.

This system has distinctly different potentials in different regions of space. This is often the case for the TISE; the prescription is to solve for the wavefunctions in each region separately, and then impose our conditions on the resulting wavefunctions.

In the case at hand, the wavefunctions cannot exist in the regions of infinite potential. Thus, we need to solve the TISE in the region $0 < x < L$, that is, the region of zero potential. Here the TISE is simply that of a free particle:

$$\frac{-\hbar^2}{2m}\frac{d^2}{dx^2}\psi(x) = E\psi(x), \qquad \text{or} \qquad \frac{d^2}{dx^2}\psi(x) = \frac{-2mE}{\hbar^2}\psi(x). \qquad (9.10)$$

It's easy to see that the general solution to Eq. (9.10) is

$$\psi(x) = A\cos\left(\sqrt{2mE/\hbar^2}\,x\right) + B\sin\left(\sqrt{2mE/\hbar^2}\,x\right), \qquad (9.11)$$

where A and B are two arbitrary constants (which must appear in the general solution to a second-order, ordinary differential equation).

Now impose continuity of the wavefunction. We know that $\psi(x)$ vanishes for $x \leq 0$ and for $x \geq L$. Thus, the solutions to Eq. (9.10) must vanish (1) at $x = 0$, and (2) at $x = L$. Condition 1 implies that we must set $A = 0$. Condition 2 implies that the argument of the sine function must satisfy

$$\sqrt{2mE_n/\hbar^2}\,L = n\pi, \qquad n = 1, 2, 3, \ldots . \qquad (9.12)$$

Here E has been replaced with E_n because there is not just *one* energy eigenvalue that satisfies condition 2, but a discrete *set* of them. Solving Eq. (9.12) for the E_ns yields

$$E_n = \frac{n^2\pi^2\hbar^2}{2mL^2}, \qquad n = 1, 2, 3, \ldots . \qquad (9.13)$$

Corresponding to the E_ns is a set of energy eigenstates, the $\psi_n(x)$s

$$\psi_n(x) = B\sin\left(\sqrt{2mE_n/\hbar^2}\,x\right) = B\sin(n\pi x/L). \qquad (9.14)$$

Note that it was the imposition of the wavefunction conditions above that resulted in "quantized" energy eigenvalues and eigenstates.

---

[8] See, for example, Liboff (2003); Schiff (1968), pp. 32–33; Townsend (1992). Vanishing of the wavefunction for infinite V may be justified by examining the wavefunctions of the *finite* potential well in the limit $V \to \infty$.

We can easily determine B from the normalization condition,

$$1 = \int_{-\infty}^{+\infty} |\psi_n(x)|^2 \, dx = B^2 \int_0^L \sin^2\left(n\pi x/L\right) dx = B^2 L/2. \qquad (9.15)$$

Thus, $B = \sqrt{2/L}$, and our normalized eigenstates are

$$\psi_n(x) = \sqrt{\frac{2}{L}} \, \sin\left(n\pi x/L\right), \qquad n = 1, 2, 3, \ldots. \qquad (9.16)$$

Though the infinite potential well is far from trivial, it is one of the simplest energy eigenvalue problems. Most other such problems are *far* more difficult. Nevertheless, the additional difficulty is largely "just" mathematics, and our primary interest lies not in applied mathematics, but in attaining a clear conceptual understanding. That is what I turn to now.

9.3 Interpretation

9.3.1 Energy Eigenstates in Position Space

The first two wavefunctions of Eq. (9.16), along with the corresponding probability densities, are plotted in Fig. (9.1). The plots provide a graphic representation of these energy eigenstates, and they're a good vehicle by which to better understand the role of the TISE and its solutions.

Let's first recall where the states of Eq. (9.16) came from. The TISE is an energy eigenvalue equation, and it holds in any representation in which we choose to work. Typically, though, we set up and solve the TISE in position representation. But this is *not* the natural representation for energy. Because this point is easily misunderstood, let's set up an analogous but simpler situation.

Suppose that for some system only two basis states are required to form a complete, orthonormal set, and that two such bases are the α basis, $|\alpha_1\rangle$ and $|\alpha_2\rangle$, and the β basis, $|\beta_1\rangle$ and $|\beta_2\rangle$. Suppose, further, that the α states

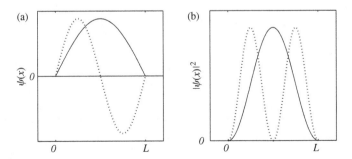

Fig. 9.1 (*a*) The $n = 1$ (solid line) and $n = 2$ (dotted line) wavefunctions for the infinite potential well, (*b*) the corresponding probability densities.

(but not the β states) are eigenstates of an operator $\hat{A}$, that is,

$$\hat{A}|\alpha_j\rangle = a_j|\alpha_j\rangle. \tag{9.17}$$

Equation (9.17) is representation-independent: though it seems most natural to work in the α basis, Eq. (9.17) must hold in any suitable basis. If we determine how $\hat{A}$ appears in the β basis, and how to expand $|\alpha_1\rangle$ and $|\alpha_2\rangle$ in terms of $|\beta_1\rangle$ and $|\beta_2\rangle$, we can write out Eq. (9.17) in the β basis, and it must hold.

Matrix mechanics clearly illustrates the difference between an eigenvalue equation when *not* in its eigenbasis, and when *in* its eigenbasis. Writing out Eq. (9.17) in matrix form for the eigenstate $|\alpha_1\rangle$, first in the β basis and then in the α basis (the eigenbasis), we have

$$\begin{pmatrix} \langle\beta_1|\hat{A}|\beta_1\rangle & \langle\beta_1|\hat{A}|\beta_2\rangle \\ \langle\beta_2|\hat{A}|\beta_1\rangle & \langle\beta_2|\hat{A}|\beta_2\rangle \end{pmatrix} \begin{pmatrix} b_1 \\ b_2 \end{pmatrix} = a_1 \begin{pmatrix} b_1 \\ b_2 \end{pmatrix}, \tag{9.18a}$$

$$\begin{pmatrix} a_1 & 0 \\ 0 & a_2 \end{pmatrix} \begin{pmatrix} 1 \\ 0 \end{pmatrix} = a_1 \begin{pmatrix} 1 \\ 0 \end{pmatrix}. \tag{9.18b}$$

Here we've taken $|\alpha_1\rangle = b_1|\beta_1\rangle + b_2|\beta_2\rangle$. Both of the above matrix equations are manifestations of Eq. (9.17), but in two different representations (bases).

When Eq. (9.17) is written in its eigenbasis—as in Eq. (9.18b)—the forms of both $\hat{A}$ and $|\alpha_1\rangle$ are simple: $\hat{A}$ has its eigenvalues on the diagonal, and zeros elsewhere, while $|\alpha_1\rangle$ consists of a 1 and a 0. But when written in the β basis—as in Eq. (9.18a)—$\hat{A}$'s diagonal elements are *not* just the eigenvalues, and $\hat{A}$ has non-zero off-diagonal elements. Moreover, both of $|\alpha_1\rangle$'s elements are now non-zero—that is, the eigenstate $|\alpha_1\rangle$, when written in the $|\beta\rangle$ basis, becomes a superposition.[9]

Returning to the infinite potential well, the TISE's solutions, Eq. (9.16), are eigenstates of $\hat{H}$, but superpositions when written in position representation. From Chapter 2 we know that position eigenstates are Dirac deltas, for example, $\delta(x - x_0)$ is the position eigenstate for the position x_0. Therefore the states of Eq. (9.16), though eigenstates of energy, are superpositions of (infinitely many) position eigenstates. (As also discussed in Chapter 2, the expansion coefficient for each position eigenstate is simply the wavefunction's value at that position.)

9.3.2 Overall and Relative Phases

Any complex function (i.e. a function with both real and imaginary parts), say $f(x)$, can be written in *polar form*: $f(x) = R(x)\exp(iS(x)/\hbar)$. Here $R(x)$ and $S(x)$ are *real* functions, called the *modulus* and the *phase*, respectively. Although the phase plays a key role in quantum mechanics, that role

---

[9] This should all sound vaguely familiar; cf. Section 6.4.

is not easily explicated. We'll discuss the phase in more depth in the next chapter, but a brief acquaintance now is worthwhile.[10]

A general rule in quantum mechanics (which I'll discuss in more detail in the next chapter) is that an overall phase doesn't alter a quantum state. This is tantamount to saying that an overall phase cannot alter any probabilities.

But just what do I mean by an *overall* phase? An overall phase is one which is the same for every component of the state. Suppose that in our example above, $b_1 = e^{iC_1}/\sqrt{2}$ and $b_2 = e^{iC_2}/\sqrt{2}$, with C_1 and C_2 real, and $C_1 \neq C_2$. Then

$$|\alpha_1\rangle = \frac{e^{iC_1}}{\sqrt{2}}|\beta_1\rangle + \frac{e^{iC_2}}{\sqrt{2}}|\beta_2\rangle, \qquad C_1 \neq C_2. \qquad (9.19)$$

Here $|\alpha_1\rangle$ exhibits a *relative* phase—one that depends on the basis state.

Now suppose, instead, that $C_1 = C_2 \equiv C$. Then

$$|\alpha_1\rangle = \frac{e^{iC}}{\sqrt{2}}\Big(|\beta_1\rangle + |\beta_2\rangle\Big) \longleftrightarrow \frac{1}{\sqrt{2}}\Big(|\beta_1\rangle + |\beta_2\rangle\Big). \qquad (9.20)$$

Because e^{iC} is an *overall* phase—the same for each basis state—it was dropped on the right-hand side. The arrow is meant to indicate that the two expressions are identical quantum states, even though they are *not* mathematically equal.

What has any of this to do with the TISE? Solutions to the TISE often are real (as for the infinite potential well), or have, at most, an overall (not relative) phase.[11] Such solutions are, effectively, *real* wavefunctions. Thus, learning quantum mechanics primarily through the TISE risks instilling misconceptions due to a focus on real wavefunctions.

Although wavefunction plots (such as Fig. 9.1) can be illustrative, they can also obscure the fact that such plots are possible only because the wavefunction is not complex. And the complex nature of quantum states, including wavefunctions, is critically important (as we'll discuss in following chapters).

Remember that the starting point in this chapter was the time-dependent Schrödinger equation. Thus, solutions to the TISE must be only part of the story. As we saw in Eq. (9.9), the full state must include a factor of $\exp(-iEt/\hbar)$.

Just what effect will this time-dependent part of the state have? The answer is: None! The factor $\exp(-iEt/\hbar)$ has no dependence on x. Thus, it is simply an overall phase for our position-space wavefunctions, and it can be dropped with impunity. It would seem, then, that the time-dependent

---

[10] Chapter 10 and Appendix A discuss the basics of complex numbers and functions.
[11] In position representation, this means the phase doesn't depend on the spatial variable(s).

part of the state is of no interest. And so long as we focus only on energy eigenstates, that's correct—and it's why many discussions of the TISE largely ignore the phase.

But what if a system is in a superposition of energy eigenstates? That's absolutely legitimate—and it can change everything. In a superposition of energy eigenstates, each eigenstate generally has a *different* energy eigenvalue, and thus a different factor $\exp(-iEt/\hbar)$, associated with it (consider the infinite potential well). This fact will play a key role in discussing time evolution in Chapter 11.

9.4 Potential Barriers and Tunneling

9.4.1 The Step Potential

Every day, in bars and pubs across the globe, "experiments" in classical scattering are conducted: whenever one billiard ball is accelerated towards, and then strikes, another, a simple classical scattering event has occurred.

Billiard balls exert forces by contact. A more interesting type of classical scattering arises from forces that extend through space—for example, two macroscopic, electrically charged objects. If, say, object 1 is very massive compared to object 2, and initially stationary, then object 1 remains essentially stationary during the scattering event. In that case, it's a good approximation to represent object 1 by an electric potential, or equivalently by an electric field, that's fixed in space. Often we can do the same thing in quantum mechanics, treating scattering as the interaction between a particle and a fixed potential. Both conceptually and calculationally, though, the quantum case is far more subtle.[12]

The simplest case of quantum scattering is that of a "particle" traveling through force-free space towards a potential step, or barrier, of height V_0 (see Figure 9.2).[13] The particle then scatters off of (interacts with) the step. This is a standard textbook problem—often found (as here) in discussions of the time-independent Schrödinger equation.[14]

But wait! Scattering is clearly a time-*de*pendent process—why discuss it in the context of time-*in*dependent quantum mechanics? First, because the step potential provides a simple application of the TISE. And because, as we'll see, step-potential scattering *can* be analyzed using the TISE, although only for a special case, and only after some rather subtle conceptual leaps have been made. (Moreover, the analysis is far simpler than using the time-dependent Schrödinger equation.)

---

[12] Our goal is to illustrate some features of quantum mechanics, not to discuss scattering *per se*—a vast subject, which most standard quantum texts treat in some depth.

[13] Speaking carefully, in quantum mechanics it's not a *particle* that's travelling, but a wavefunction—a quantum state—from which we may extract measurement probabilities.

[14] Many "modern physics" texts use the TISE to discuss both step-potentials and tunneling.

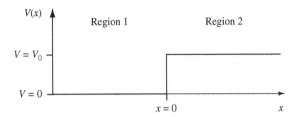

Fig. 9.2 The step potential; for Region 1, $x < 0$, while for Region 2, $x > 0$.

Before discussing the step potential as a scattering problem, however, we'll examine it as an energy eigenvalue problem. The Hamiltonian operator for the step potential is:

$$\hat{H} = \frac{\hat{p}^2}{2m} = \frac{-\hbar^2}{2m}\frac{d^2}{dx^2} \qquad \text{in Region 1,}$$

$$\hat{H} = \frac{\hat{p}^2}{2m} + V_0 = \frac{-\hbar^2}{2m}\frac{d^2}{dx^2} + V_0 \qquad \text{in Region 2.} \qquad (9.21)$$

In both regions, solutions to the TISE are of the same form: $\Psi_1(x) = A_+e^{ik_1x} + A_-e^{-ik_1x}$, and $\Psi_2(x) = B_+e^{ik_2x} + B_-e^{-ik_2x}$ (1 and 2 denote regions; A_+, A_-, B_+, and B_- are complex constants). Substituting these into the TISE easily yields the system energy, E:

$$E = \frac{\hbar^2 k_1^2}{2m} = \frac{\hbar^2 k_2^2}{2m} + V_0. \qquad (9.22)$$

Thus, k_1 and k_2 satisfy:

$$k_1 = \frac{\sqrt{2mE}}{\hbar}, \qquad k_2 = \frac{\sqrt{2m(E - V_0)}}{\hbar}. \qquad (9.23)$$

If $E > V_0$, then both k_1 and k_2 are positive, real numbers, and $k_1 > k_2$. Moreover, from $e^{\pm i\alpha x} = \cos(\alpha x) \pm i\sin(\alpha x)$ (see Section 10.1.2), we see that both the real and imaginary parts of $\Psi_1(x)$ and $\Psi_2(x)$ are oscillatory (though with different frequencies for $\Psi_1(x)$ and $\Psi_2(x)$).

But what if $E < V_0$? Then k_1 remains real and positive, but because $\sqrt{2m(E - V_0)}$ is now the square root of a *negative* number, k_2 becomes *imaginary*: $k_2 = \pm i\sqrt{2m(V_0 - E)}/\hbar$. Thus, the exponential in $\Psi_2(x)$ becomes *real*: $\Psi_2(x) = C_+e^{\kappa x} + C_-e^{-\kappa x}$, where $\kappa \equiv \sqrt{2m(V_0 - E)}/\hbar$. Though fine mathematically, this seems physically absurd: How can the total energy, E, be less than the potential energy, V_0? Good question: but for now, I'll defer answering it.

Evidently we have two distinct cases: $E > V_0$ and $E < V_0$. For the $E > V_0$ case, the most general energy eigenstate would include all four possible constituent states: $\Psi_1(x) = A_+e^{ik_1x} + A_-e^{-ik_1x}$, and $\Psi_2(x) = B_+e^{ik_2x} + B_-e^{-ik_2x}$.

For the $E < V_0$ case, also, there appear to be four possible states: $\Psi_1(x) = A_+ e^{ik_1 x} + A_- e^{-ik_1 x}$, and $\Psi_2(x) = C_+ e^{\kappa x} + C_- e^{-\kappa x}$. However, because κ is a positive real number, $\Psi_2(x) = C_+ e^{+\kappa x}$ diverges (approaches infinity) as $x \to \infty$, and so is not an acceptable wavefunction.

We could now finish constructing our energy eigenstate wavefunctions for each case by imposing continuity of the wavefunction and its first derivative at the boundary between the regions, that is, at $x = 0$.[15] This is a fairly straightforward exercise—one I'll leave to the standard textbooks.[16]

9.4.2   The Step Potential and Scattering

So far, we've treated the step potential as an energy eigenvalue problem. That is, we've applied the TISE to the step potential to obtain the energy eigenvalues and eigenstates (in position representation). But how is this relevant to the time-*de*pendent process of scattering?

To correctly model scattering, the initial state must be a wavepacket—a state with finite spread in both position and momentum (e.g., a Gaussian)—which is time evolved until interaction with the potential effectively ends. Wavepackets must obey the position-momentum uncertainty relation: $\Delta x \Delta p_x \geq \hbar/2$. This means, roughly, that the price to be paid for a narrow momentum probability distribution is a wide position probability distribution, and vice versa.

In the limit of an infinitesimally narrow momentum distribution—a momentum eigenstate—the corresponding position distribution is maximally spread out. Such a state is physically unrealizable, but states with well (but not perfectly) defined momenta *are* realizable. A momentum eigenstate is a good approximation to such a state.

In Region 1, the step potential's energy eigenstate is $\Psi_1(x)$ (see Section 9.4.1). Both "pieces" of $\Psi_1(x)$, that is, $A_+ e^{ik_1 x}$ and $A_- e^{-ik_1 x}$, are spread out over all of Region 1, and each has a well-defined momentum (to check, let $\hat{p} = -i\hbar\, d/dx$ act on each). Thus, we can think of $A_+ e^{ik_1 x}$ as representing an ensemble of particles, all with momentum $p = +\hbar k_1$ (and similarly for $A_- e^{-ik_1 x}$, but with $p = -\hbar k_1$). For the $E > V_0$ case, similar arguments apply in Region 2. Thus, if we solve the time *in*dependent Schrödinger equation for the step potential's energy eigenstates, we've obtained a sort of "steady-state solution" to the time-*de*pendent scattering problem.

Consider the $E > V_0$ case in detail. Because we're modelling the interaction of an incoming beam of particles with a potential step, we can impose an additional condition on our wavefunction. Physically, we have a beam of particles incident from the left, but no such beam is incident from the right.

---

[15] For the step potential, V *is* discontinuous (as for the infinite potential well), but it is *not* infinite, so continuity of the derivative applies in this case. See note 6.

[16] See, for example, Liboff (2003), Section 7.6; Shankar (1980), Section 5.4; Townsend (1992), Section 6.10.

(a) $E > V_0$ (b) $E < V_0$

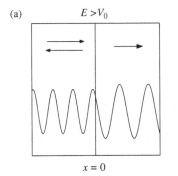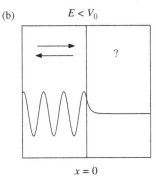

$x = 0$ $x = 0$

Fig. 9.3 Real parts of the (complex) scattering wavefunctions for the step potential, for (*a*) $E > V_0$, and (*b*) $E < V_0$. (The imaginary parts are similar.) Arrows indicate the momenta in the two regions. Such representations are idealized: physically realistic states would be localized wavepackets, not infinitely extended, time-independent waves.

Therefore, although $B_- e^{-ik_2 x}$ is mathematically acceptable in $\Psi_2(x)$, it's not physically acceptable, and must vanish (i.e., we set $B_- = 0$).

Now let's cheat just a bit by thinking in terms of particles. In Region 1, incident particles move in the $+x$ direction with kinetic energy E; transmitted particles, in Region 2, also move in the $+x$ direction, but with kinetic energy $E - V_0$.[17] So far this seems quite classical. However, even though $E > V_0$, some of the incident particles are *reflected* at $x = 0$, traveling in the $-x$ direction with energy E.[18] This effect is clearly *non*-classical—and somewhat surprising. (See Fig. 9.3 (*a*).)

For the $E < V_0$ case, things get more interesting. A classical particle with $E < V_0$ could never enter Region 2. But the wavefunction does. What does this mean for the quantum particle?

First, even the question is dangerously misleading. In quantum mechanics, it's not a *particle* that enters Region 2, but a *wavefunction—not* the same thing! Because the state has non-zero amplitude in Region 2, there's a non-zero probability of finding the particle there upon measurement. But how do we "think about" the particle in Region 2? That's not clear, and it's why a question mark appears in Fig. 9.3 (*b*).

One problem with the conceptual picture we've formed is that we only suspend our classical judgement when convenient. In Region 2, a particle beam moving to the left is eliminated because, classically, no such beam

---

[17] This reduction in kinetic energy, and thus speed, occurs *instantaneously*. This physically impossible effect arises from the physically unrealistic discontinuous change in potential.
[18] Recall that, in Region 1, the wavefunction is comprised of both $A_+ e^{ik_1 x}$ and $A_- e^{-ik_1 x}$.

exists. Yet for the $E > V_0$ case, a reflected beam is retained in Region 1, even though reflection is prohibited classically. Similarly, for the $E < V_0$ case, the wavefunction enters Region 2, even though a classical particle can't.

Arguably, such conceptual questions result from the fact that we are solving the problem as an approximation to, and idealization of, the real situation. The proper way to gain a clear understanding of the physics, and thus justify our conceptual picture, is by examining the real, wavepacket-based scattering problem. I'll return to this topic briefly in Section 12.3.2.

9.4.3   Tunneling

Because in Section 9.4.2 we obtained a time-independent state—quite different from the time-dependent wavepackets of the "real" problem—it may seem that we've done little of physical interest. But we can still address a key physical question: For some incident particle beam, what is T, the fraction of particles transmitted, and R, the fraction reflected?

The process of matching boundary conditions at $x = 0$ (which I discussed, but didn't actually carry out, in Section 9.4.1) determines the relative sizes of the constants A_+, A_-, B_+ (for $E > V_0$) and C_- (for $E < V_0$). The ratios $|A_-|^2/|A_+|^2$ and $|B_+|^2/|A_+|^2$ are the probability densities for the reflected and transmitted beams, respectively, relative to the incident beam.[19] To obtain T and R, however, these ratios must be appropriately weighted by the speeds of the particles in the two beams (I again leave details to the comprehensive texts).[20]

Now consider a somewhat different potential: a step potential of finite length, say L, returning to $V = 0$ for $x > L$ (see Fig. 9.4). For $x < 0$ and for $x > L$, the wavefunction is similar to that for the step potential, that is, $A_+e^{ik_1x} + A_-e^{-ik_1x}$ for $x < 0$, and $B_+e^{ik_1x}$ for $x > L$. Again invoking a particle picture, for $x < 0$ and $x > L$, particles move with kinetic energy E.

What happens in the barrier region? For the case $E > V_0$, a particle's kinetic energy is reduced to $E - V_0$. As for the step potential, some of the

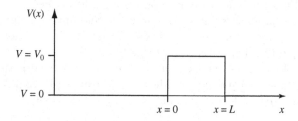

Fig. 9.4   The rectangular potential barrier.

---

[19] There is no transmitted beam for the $E < V_0$ case.

[20] Rigorously, this is accounted for with the *probability current density*, or *probability flux*.

particles are reflected at $x = 0$. In addition, another reflection can occur at $x = L$. Apart from these non-classical reflections, however, the overall picture agrees fairly well with our classical expectations.

When we consider the $E < V_0$ case, however, things go from interesting to profound. Because the potential barrier is of finite extent, the wavefunction in the barrier, though rapidly decreasing, is non-zero at L.

What happens at L? Again, continuity of the wavefunction and its first derivative must hold. This means, in part, that because the wavefunction within the barrier is non-zero at L, so must there be a non-zero wavefunction for $x > L$. But this means that there is a non-zero probability that particles pass through to the other side of the potential barrier, where they are again free particles.

Using techniques like those applied to the step potential, we can again arrive at transmission and reflection probabilities. But our real interest is in the conceptual analysis of the situation.

For both the step potential and the rectangular potential, the physical meaning of the wavefunction's penetration into the forbidden region is unclear. For the rectangular potential, however, the end result of that penetration *is* clear—and quite astounding. Some of the particles incident from the left are reflected from the barrier. But others pass through the barrier, a region where they could not even *exist* classically, and reappear on the other side! This is quantum-mechanical *tunneling*, and it has no counterpart in classical physics.

By using the TISE to introduce tunneling, we avoided much of the mathematics that arises in a rigorous treatment using wavepackets and the time-dependent Schrödinger equation. But, as for the step potential, the conceptual aspects are perhaps more clear in the latter context.[21]

9.5   What's Wrong with This Picture?

Solving the time-independent Schrödinger equation constitutes a major part of quantum physics, and we've only scratched the surface. Important as the TISE is, though, its study can instill misconceptions, so let's sum up how those misconceptions may take root.

Newton's second law, $\vec{F}_{net} = m\vec{a}$, describes a wealth of classical physics. An important subset of classical physics comprises systems in static equilibrium. For such systems the second law reduces to $\vec{F}_{net} = 0$.

Although the acceleration now seems to have disappeared from the second law, it really hasn't—it's just that now $\vec{a} = 0$. Still, if we focused on static equilibrium problems we might lose sight of the centrality of acceleration in classical mechanics, since it would play no meaningful role in either our concepts or our calculations.

---

[21] See, for example, Goldberg (1967); Schiff (1968), pp. 106–107.

The time-independent Schrödinger equation describes a wealth of quantum physics. Energy eigenstates and eigenvalues play a crucial role in our understanding of the physical world, especially in nuclear, atomic, and molecular physics and chemistry. But for any individual energy eigenstate, time evolution amounts to nothing more than multiplication by an overall phase. In a sense, energy eigenstates *do* evolve in time, but in such a way that there are *no* physical consequences.

The classical world doesn't consist only of systems in static equilibrium. And the quantum world doesn't consist only of energy eigenstates; rather, it is full of systems that *do* change in time, and we can't understand these by studying the TISE alone.

Precisely how can the TISE cause us problems down the road? First, solutions to the TISE (in position space) are often *real* wavefunctions, which can't fully reflect the richness of quantum mechanics. In general, wavefunctions are complex. Unlike real functions, they simply can't be drawn on paper.[22]

Moreover, even when solutions to the TISE are complex, that fact often plays little role in introductory treatments. After all, if we're just going to complex square the state and obtain a *real* probability distribution, isn't the complex nature of quantum states simply an unnecessary mathematical artifact?

Finally, much interesting quantum mechanics arises from *superpositions* of energy eigenstates. And dealing with such superpositions requires us to deal explicitly with the complex nature of the wavefunction. Quantum interference, which gives rise to many uniquely quantum-mechanical phenomena, is crucially dependent on the complex nature of the state.

In the next chapter, we will explore in depth the role of the phase, and the complex nature of the state.

9.6 Problems

1. Consider the $n = 2$ energy eigenstate of the infinite potential well. Suppose that, rather than adopt the statistical interpretation, we interpret this state as describing an actual, physical particle with a well-defined energy (which must all be kinetic). Carefully describe how the energy and position probability distributions (i) are, and/or (ii) are not in agreement with such an interpretation.

2. Consider the following statement. For each sentence, state whether you agree with it, and justify your answer.

   > Solutions to the TISE are energy eigenstates. These solutions are functions of position, x. Therefore, there is a unique energy corresponding to each value of x for such a state.

---

[22] Of course, one *can* plot, say, the real part of a wavefunction, or the imaginary part, or the modulus. But these are not the wavefunction itself.

3. The potential energy of a classical harmonic oscillator is: $V = -\kappa x^2$, where κ is a positive, real constant.
   (i) Set up (but don't solve) the TISE for the corresponding quantum harmonic oscillator.
   (ii) Now assume there's also a damping force of $(-\alpha/m)p$, where m is the mass, α is a real constant, and p is the momentum. Repeat Part (i) for this case.

4. For a rectangular potential barrier, the methods of Section 9.4 predict that T, the fraction of particles transmitted, will be

$$T = \left[1 + \frac{V_0^2 \sinh^2(\kappa L)}{4E(V_0 - E)}\right]^{-1} \qquad E < V_0,$$

$$T = \left[1 + \frac{V_0^2 \sin^2(k_2 L)}{4E(E - V_0)}\right]^{-1} \qquad E > V_0, \qquad (9.24)$$

   where κ, k_2, and L are defined in Section 9.4. To investigate T's behavior, set $m = \hbar = 1$, take $V_0 = 10$, and plot T as function of:
   (i) energy, with $0 \le E \le 50$,
   (ii) barrier width, with $E = 20$ and $0 \le L \le 2$,
   (iii) barrier width, with $E = 5$ and $0 \le L \le 2$.

5. Computer-generated "movies" of wavepacket scattering from a potential barrier, including tunneling, can be helpful in forming physical pictures of such processes. Using Goldberg (1967), Schiff (1968) pp. 106–107, or other sources, investigate wavepacket scattering from a rectangular potential barrier.
   (i) In discussing both the step and rectangular potential barriers, we considered the $E > V_0$ case, and referred to particles with kinetic energy E, or $E - V_0$. Is this language applicable for wavepackets? That is, does a wavepacket have a well defined energy? Justify your answer.
   (ii) Consider the TISE results for the step and rectangular barriers for the $E > V_0$ case. Compare these results with wavepacket scattering. In what ways does the TISE solution accurately portray what happens with wavepackets? In what ways is it misleading or incomplete?
   (iii) Repeat Part (i) for the $E < V_0$ case.

10

Why Is the State Complex?

Born advanced [a theorem] which [was] destined to play a fundamental role in the further development of quantum theory, its interpretation, and its theory of measurement... the theorem of the interference of probabilities according to which the phases of the expansion coefficients, and not only their absolute values, are physically significant.[1]

It was 1926—the dawn of quantum mechanics—when Max Born realized that quantum-mechanical expansion coefficients, and thus quantum states themselves, cannot simply be real quantities: they *must*, in general, be complex.

But why? In one sense, this question is easily answered. The Hamiltonian operator, $\hat{H}$, is real,[2] so solutions to the time-dependent Schrödinger equation,

$$i\hbar\frac{d\Psi}{dt} = \hat{H}\Psi, \tag{10.1}$$

must be complex—as can be seen by assuming the contrary, that is, that Ψ is *only* real or *only* imaginary. This does explain, formally, why the state is complex, but we would like to know more. What essential features of quantum mechanics depend on the state's complex nature? How and when do these features arise?

Such questions are often neglected, yet one can't fully understand quantum mechanics until they've been addressed. To do so, we'll need a clear understanding of certain aspects of complex numbers, and developing that understanding will be our first goal. With that in hand, we'll investigate the role of the phase in quantum mechanics. We'll first see that the full range of physical possibilities can only be incorporated into complex states. Then we'll develop the difference between overall and relative phases, the effects of each on probabilities, and how such phases arise through unitary operators.

---

[1] Jammer (1966), p. 290.
[2] $\hat{H}$ *can* involve i, but this is not the usual case. See Schiff (1968), Section 20.

10.1 Complex Numbers

10.1.1 Basics

Grasping the implications of complex quantum states requires a basic but firm understanding of complex numbers. To that end, this section deals not with physics *per se*, but with mathematics. Although discussions of complex numbers are easily found elsewhere, the approach here is specifically designed to lay the groundwork for our later foray into the complex nature of quantum states.[3]

The starting point for complex numbers is, of course, the imaginary number i, where $i \equiv \sqrt{-1}$. If C_1 is a complex number, then by definition C_1 has both a real part, denoted Re C_1, and an imaginary part, denoted Im C_1. Both Re C_1 and Im C_1 are *real* numbers; let's agree to call them α and β, respectively. Then we can write C_1 as

$$C_1 = \text{Re } C_1 + i\text{Im } C_1 = \alpha + i\beta. \tag{10.2}$$

I'll call this the "standard form" for a complex number.

Addition, subtraction, and multiplication of complex numbers are defined in the "obvious" way, that is, as they would be for real numbers. Suppose another complex number, C_2, is given by $C_2 = \gamma + i\delta$. Then

$$C_1 + C_2 = \alpha + \gamma + i(\beta + \delta)$$
$$C_1 - C_2 = \alpha - \gamma + i(\beta - \delta)$$
$$C_1 C_2 = (\alpha + i\beta)(\gamma + i\delta) = \alpha\gamma - \beta\delta + i(\alpha\delta + \beta\gamma). \tag{10.3}$$

It's crucially important to recognize that real and imaginary quantities are like "apples and oranges"; they are fundamentally different. For example, if $C_1 = C_2$, then we must have $\alpha = \gamma$ *and* $\beta = \delta$; the real and imaginary parts of the sum must *each* be equal. Thus, $C_1 = C_2$ is effectively *two* equations.

By definition, the *complex conjugate* (or just "conjugate") of some quantity D, denoted D^*, is obtained by simply changing the sign of i everywhere it appears in D. For a complex number such as C_1, then, $C_1^* = \alpha - i\beta$.

Moreover, $|D|^2$ is defined to mean "multiply D by D^*, its complex conjugate." So for $|C_1|^2$, we have

$$|C_1|^2 = C_1 C_1^* = (\alpha + i\beta)(\alpha - i\beta) = \alpha^2 + \beta^2. \tag{10.4}$$

Simple. But there's a potential problem. When dealing with complex quantities, one must be scrupulously careful to distinguish ordinary multiplication, such as in Eq. (10.3), from multiplication of a quantity by its

---

[3] This section is intended for those with at least some familiarity with complex numbers. If you have none, I suggest you consult a more systematic treatment. Also see Appendix A.

conjugate, such as in Eq. (10.4). Both operations are legitimate, and both could appear in a calculation—but they obviously don't yield the same results, for example, $C_1^2 \neq |C_1|^2$.

10.1.2 Polar Form

Any complex number can be written in standard form, that is, in terms of its real and imaginary parts (as in Eq. (10.2)). Often, however, it's more convenient to use an alternative form, called *polar* form.

An ordinary vector $\vec{A}$ in the xy plane (Fig. 10.1(a)) can be fully specified with just two pieces of information, such as A_x and A_y, its x and y components. Instead, however, we could specify $\vec{A}$'s length and its angle with respect to the x axis. Given A_x and A_y, we can find the length and angle of $\vec{A}$, and vice versa. Which of these two means of specifying $\vec{A}$ is more convenient depends on context.

The *complex plane* is a plane consisting of a *real* axis, and an orthogonal *imaginary* axis—corresponding to the real and imaginary parts of a complex number. We can represent a complex number as a vector in this plane (see Fig. 10.1(b)).[4] Clearly, the complex plane is an abstract mathematical creation. Unlike the xy plane, we can't think of the complex plane as representing physical space. What's important, however, is that the same rules apply for vectors (complex numbers) in the complex plane as apply to ordinary vectors in the xy plane. Let's see how.

Writing the ordinary vector $\vec{A}$ as $\vec{A} = A_x\hat{x} + A_y\hat{y}$ specifies $\vec{A}$'s components. Similarly, writing a complex number in standard form, as in Eq. (10.2), corresponds to specifying the number's components. In Figure 10.1(b), α and β are the components of C_1.

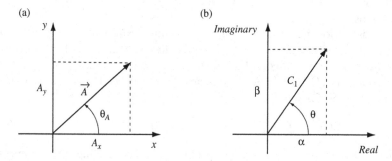

Fig. 10.1 The analogy between (a) a vector, $\vec{A}$, in the xy plane, and (b) a complex number, C_1, in the complex plane.

---

[4] This *doesn't* mean that one axis exists in reality and the other only in one's imagination.

In Figure 10.1(b), we see that the "length", properly called the *modulus*, of C_1 is $\sqrt{\text{Re } C_1^2 + \text{Im } C_1^2} = \sqrt{\alpha^2 + \beta^2}$. And just as the angle of $\vec{A}$ is $\tan^{-1}(A_y/A_x)$, the "angle", properly called the *phase*, of C_1 is $\tan^{-1}(\text{Im } C_1/\text{Re } C_1)$. By convention, the length or modulus is denoted R, and the angle is denoted θ.

Again referring to Fig. 10.1(b), we could write C_1 in component form as $C_1 = R(\cos\theta + i\sin\theta)$. Using *Euler's formula*,[5] $\exp(\pm ix) = \cos(x) \pm i\sin(x)$, then, we have, $C_1 = R\exp(i\theta)$. A complex number written in this manner—as the product of a *real* modulus, R, and a complex phase factor (with a *real* phase, θ)—is said to be in *polar form*.

Polar form can be very convenient if a complex number's modulus or phase are of interest, since they can then simply be "read off". But beware! It's easy to wrongly think that the modulus R and the phase factor $e^{i\theta}$ are the real and imaginary parts, respectively, of a complex number. After all, that's what they *look* like—until you write out the exponential and obtain $R\exp(i\theta) = R(\cos\theta + i\sin\theta)$.

By contrast, the modulus and phase of a complex number written in standard form *can't* simply be read off, but its real and imaginary parts *can*. Again, which form is preferable depends on context.

10.1.3 Argand Diagrams and the Role of the Phase

Although complex numbers aren't really vectors, we can *represent* them as such, as in Fig. 10.1(b). Such representations, called *Argand diagrams*, provide a useful means of visualization.

The analogy between ordinary vectors and complex numbers extends to addition. Consider the sum of $\vec{A}$ and $\vec{B}$, two vectors in the xy plane:

$$\vec{A} + \vec{B} = (A_x\hat{x} + A_y\hat{y}) + (B_x\hat{x} + B_y\hat{y})$$
$$= (A_x + B_x)\hat{x} + (A_y + B_y)\hat{y}. \qquad (10.5)$$

Just as we added $\vec{A}$ and $\vec{B}$ by x and y components, so do we add complex numbers by real and imaginary components. For $C_1 + C_2$ we have

$$C_1 + C_2 = (\text{Re } C_1 + i\text{Im } C_1) + (\text{Re } C_2 + i\text{Im } C_2)$$
$$= (\text{Re } C_1 + \text{Re } C_2) + i(\text{Im } C_1 + \text{Im } C_2). \qquad (10.6)$$

(Compare with Eq. (10.3).) This implies that, just as we can graphically add ordinary vectors in the xy plane, we can graphically add complex numbers in the complex plane. By putting this principle to work in an example, we can begin to see the crucial role of the phase.

First, let's again consider some complex number C, written in polar form as $C = R\exp(i\theta)$. If we represent C as a vector in the complex plane

---

[5] This is an important identity. Be sure you are familiar with it.

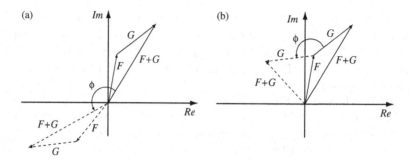

Fig. 10.2 Addition of complex numbers. (*a*) The moduli of F, G, and $F+G$ are unaltered by an overall phase change. (*b*) A relative phase change alters the modulus of $F + G$. Dashed lines represent quantities after phase shift.

using an Argand diagram, the modulus R is the vector's length, while the phase θ is the vector's angle. What if we were to multiply C by, say, $\exp(i\pi/4)$, thus altering its phase? That is,

$$C = R\exp(i\theta) \longrightarrow R\exp(i\pi/4)\exp(i\theta) = R\exp[i(\pi/4 + \theta)]. \quad (10.7)$$

Clearly, this changes only the *angle* of the vector, not its modulus.

What if we were to *add* two complex numbers? Let's call the two complex numbers F and G, and let's assume that they have *equal* moduli, R, but *unequal* phases, θ_F and θ_G. Figure 10.2(*a*) shows $F+G$ graphically in the complex plane. If we now "shift" the phases of *both* F and G by the *same* amount ϕ, we have,

$$Re^{i\theta_F} + Re^{i\theta_G} \longrightarrow Re^{i(\theta_F+\phi)} + Re^{i(\theta_G+\phi)}. \quad (10.8)$$

This *overall* phase shift (i.e. the same for both F and G) changes the orientation of both vectors equally, so that their *relative* orientation is unaltered. As a result, the modulus of the sum, $F+G$, is also unaltered by this overall phase change (Fig. 10.2(*a*)).

So, what's the effect of a *relative* phase change—when the phase change for F differs from that for G? To answer this, it suffices to change the phase of just one of our complex numbers, say G:

$$Re^{i\theta_F} + Re^{i\theta_G} \longrightarrow Re^{i\theta_F} + Re^{i(\theta_G+\phi)}. \quad (10.9)$$

Clearly, G now rotates in the complex plane, while F does not—and this changes everything. Now the modulus of $F + G$ *does* change, even though the moduli of F and G individually are unaltered (Fig. 10.2(*b*)). It's the phase that effected this for us: in fact, because we've taken the moduli of F and G to be equal, there's some ϕ for which $F + G$ vanishes identically! This illustrates the crucial role of relative phase changes.

Some remarks about terminology are in order. In physics, words some-times have completely different meanings depending on context; *phase* is one such word. In thermodynamics, phase refers to a state of matter, such as solid or liquid. In classical Hamiltonian mechanics, we work in an abstract mathematical space called *phase space*. In astronomy we refer to the phases of the moon. And the phase of a classical wave, such as a sound wave or a wave on a string, is the wave's relative spatial position. For example, $\sin(x - \pi)$ is identical to $\sin(x)$, apart from a shift in the $+x$ direction by π due to the phase, $-\pi$.

These meanings of "phase" seem wholly unrelated to the phase of a com-plex number, but that's not quite true. If we add (superpose) two classical waves, their phases partly determine the resulting wave. For example, if we add two similar sine waves, the sum depends on their phases. If the waves' peaks "line up," as for $\sin(x) + \sin(x)$, the sum is another sine wave, but with doubled amplitude. But if one wave's peaks line up with the other's troughs, as for $\sin(x) + \sin(x - \pi)$, the waves are exactly out of phase, and their sum vanishes identically. We've now seen that the phases of complex numbers determine how they add, so there's at least a rough similarity between the phase of a complex number and that of a classical wave.[6]

We've discussed complex numbers in some depth now. But quantum mechanics uses more than just complex *numbers*; as we emphasized in Chapter 9, it uses complex *functions*, which we've ignored so far in this chapter. However, almost all of our discussion about complex numbers translates to complex functions—a point that we'll explore later.

10.2 The Phase in Quantum Mechanics

10.2.1 Phases and the Description of States

In what follows, I'll illustrate the role of complex numbers in quan-tum mechanics. The underlying principle will be that, because quantum mechanics makes probabilistic predictions, only those changes to a quantum state that alter probabilities have physical significance—if no probabilities are altered, we really have the *same* state.

Let's consider the simplest case possible: a system for which only two states are required to form a complete, orthonormal set. A spin $\frac{1}{2}$ system is of just this type. As we saw in Chapter 8, a spin $\frac{1}{2}$ basis set is formed by the states $|1/2, 1/2\rangle$ and $|1/2, -1/2\rangle$; that is, $l = 1/2$ and $m = 1/2, -1/2$. If it's understood that we're working with a spin $\frac{1}{2}$ system (in the z basis), it's common to use a simpler notation, in which we replace $|1/2, 1/2\rangle$ and $|1/2, -1/2\rangle$ with $|+z\rangle$ and $|-z\rangle$, respectively. Because $|+z\rangle$ and $|-z\rangle$ form a basis, an arbitrary state of our spin $\frac{1}{2}$ system can be written as $c_+|+z\rangle + c_-|-z\rangle$, where c_+ and c_- are *complex* numbers.

---

[6] Beware that both this "wave phase" and the phase of a complex number can appear in quantum mechanics—and both are often simply (and ambiguously) called "the phase."

The z component of the spin angular momentum, denoted S_z, is an observable quantity. For a spin $\frac{1}{2}$ system, the corresponding Hermitian operator, $\hat{S}_z$, satisfies the eigenvalue equation

$$\hat{S}_z | \pm z \rangle = \pm \tfrac{\hbar}{2} | \pm z \rangle, \tag{10.10}$$

so the possible measured values for S_z are $+\frac{\hbar}{2}$ and $-\frac{\hbar}{2}$. Similarly, for the x and y components of spin angular momentum we have

$$\hat{S}_x | \pm x \rangle = \pm \tfrac{\hbar}{2} | \pm x \rangle, \qquad \hat{S}_y | \pm y \rangle = \pm \tfrac{\hbar}{2} | \pm y \rangle. \tag{10.11}$$

To even write down a quantum state we must, in general, allow it to be complex (i.e. to have complex expansion coefficients). Otherwise, we cannot account for all experimental possibilities. As a simple example, consider two spin $\frac{1}{2}$ states, $|A\rangle$ and $|B\rangle$, written in the z basis:

$$|A\rangle = \tfrac{1}{\sqrt{2}} | + z \rangle + \tfrac{1}{\sqrt{2}} | - z \rangle, \qquad |B\rangle = \tfrac{1}{\sqrt{2}} | + z \rangle - \tfrac{1}{\sqrt{2}} | - z \rangle. \tag{10.12}$$

Clearly, if we measure S_z with the system in either $|A\rangle$ or $|B\rangle$, we have

$$Prob\left(S_z = \tfrac{\hbar}{2}\right) = \tfrac{1}{2}, \qquad Prob\left(S_z = \tfrac{-\hbar}{2}\right) = \tfrac{1}{2}. \tag{10.13}$$

That is, $|A\rangle$ and $|B\rangle$ are indistinguishable with respect to S_z measurements.

Yet $|A\rangle$ and $|B\rangle$ are definitely *not* the same state. It turns out that transforming $|A\rangle$ and $|B\rangle$ into the x basis yields

$$|A\rangle = | + x \rangle, \qquad |B\rangle = | - x \rangle. \tag{10.14}$$

So if we measure S_x, rather than S_z, on $|A\rangle$ and on $|B\rangle$, we obtain

$$Prob\left(S_x = \tfrac{\hbar}{2} \big| A\right) = 1, \qquad Prob\left(S_x = \tfrac{-\hbar}{2} \big| A\right) = 0$$
$$Prob\left(S_x = \tfrac{\hbar}{2} \big| B\right) = 0, \qquad Prob\left(S_x = \tfrac{-\hbar}{2} \big| B\right) = 1. \tag{10.15}$$

The probabilities for S_x measurements for $|A\rangle$ are *opposite* those for $|B\rangle$! The complex nature of the state is what's done this for us.

But wait! you protest. The coefficients of $|A\rangle$ and $|B\rangle$ in the z basis were *real*—there was nothing complex, or even imaginary, in sight. Right; and wrong. The two states differ only by a sign change in the coefficient of $|-z\rangle$. But -1 can be written as a complex number: $-1 = e^{i\pi} = \cos(\pi) + i\sin(\pi)$, so that

$$|B\rangle = \tfrac{1}{\sqrt{2}} | + z \rangle + \tfrac{e^{i\pi}}{\sqrt{2}} | - z \rangle. \tag{10.16}$$

Thus, we may regard our sign change as a *phase change* in the coefficient of $|-z\rangle$.

Perhaps this seems like a ruse. But consider another state:

$$|C\rangle = \tfrac{1}{\sqrt{2}} | + z \rangle + \tfrac{i}{\sqrt{2}} | - z \rangle. \tag{10.17}$$

The state $|C\rangle$ is, in fact, $|+y\rangle$ in disguise—and distinct from either $|+x\rangle$ or $|-x\rangle$.[7] The probabilities for S_z measurements in $|C\rangle$ are the same as in $|A\rangle$ or $|B\rangle$ (see Eq. (10.13)), but the obviously *unreal* coefficient of $|-z\rangle$ in Eq. (10.17) leads to different probabilities for other measurements.[8]

In fact, infinitely many *distinct* spin $\frac{1}{2}$ states exist that obey Eq. (10.13). All such states must have $|+z\rangle$ and $|-z\rangle$ coefficients of modulus $\frac{1}{\sqrt{2}}$, so as to satisfy Eq. (10.13), yet all must be *different* quantum states. How can we possibly satisfy these two requirements? The answer is: by multiplying the modulus of one of the coefficients by a complex phase factor e^{ig}, where g is a different real number for each state (which is just what differentiates $|A\rangle$, $|B\rangle$, and $|C\rangle$). Evidently what distinguishes these states from each other is *not* the coefficients' moduli, but their *phases*.

Often in quantum mechanics we may deal with states, such as $|A\rangle$ and $B\rangle$, whose coefficients are real. But real coefficients can always be written as complex numbers. A real number N can always be written as Ne^{i0}, and $-N$ as $Ne^{i\pi}$.

In fact, for consistency, and to impart a deeper understanding of quantum states, one could think of *all* quantum-mechanical coefficients as complex numbers. In some cases, however, the imaginary part of a coefficient will be zero, so it *may* be written as a real number.

The situation is similar for wavefunctions. There are infinitely many wavefunctions with identical moduli—and thus identical position probability distributions—but different phases. These are thus different states, with different momentum probability distributions (as is discussed in Chapter 12).

Evidently the phase plays a crucial role in quantum mechanics. And phases can not only *exist* in the description of a quantum state, they can be *introduced*—as we'll soon see.

10.2.2   Phase Changes and Probabilities

We now consider three types of changes to the coefficients of our arbitrary spin $\frac{1}{2}$ state, $c_+|+z\rangle+c_-|-z\rangle$, and ask whether they can alter probabilities.

An *overall* phase change of, say, $e^{i\alpha}$, with α constant, leads to

$$c_+|+z\rangle + c_-|-z\rangle \longrightarrow e^{i\alpha}c_+|+z\rangle + e^{i\alpha}c_-|-z\rangle. \qquad (10.18)$$

Note that: $|c_+|^2 = |e^{i\alpha}c_+|^2$, and $|c_-|^2 = |e^{i\alpha}c_-|^2$. That is, the overall phase change leaves the (complex) squares of the coefficients, and thus the measurement probabilities for S_z, unaltered.

---

[7] Writing $i = e^{i\pi/2}$, we again see that a *phase change* created a new state; in this case, $|C\rangle$.

[8] For example, S_x measurements in $|C\rangle$ satisfy: $Prob(S_x = \frac{\hbar}{2}) = Prob(S_x = \frac{-\hbar}{2}) = \frac{1}{2}$.

A *relative* phase change may be effected by changing just one of the phases

$$c_+| + z\rangle + c_-| - z\rangle \longrightarrow e^{i\alpha}c_+| + z\rangle + c_-| - z\rangle. \qquad (10.19)$$

We might expect that the probabilities for S_z would be altered by this relative phase change. Clearly, though, the complex squares of the coefficients of $|+z\rangle$ and $|-z\rangle$, and thus also the S_z probabilities, are unaltered. As we'll soon see, this does *not* contradict the dictum that "overall phase changes don't change states, but relative phase changes do."

Our final case might at first seem a bit odd. First write one of the coefficients, say c_+, as the sum of two other complex numbers: $c_+ = n_1 + n_2$. Calculating $|c_+|^2$ in terms of n_1 and n_2, we have

$$|c_+|^2 = |n_1 + n_2|^2 = |n_1|^2 + |n_2|^2 + n_1 n_2^* + n_2 n_1^*$$
$$= |n_1|^2 + |n_2|^2 + 2\mathrm{Re}(n_1 n_2^*). \quad (10.20)$$

Now suppose that n_1 somehow "picks up" a phase of $e^{i\sigma}$ (with σ constant). Clearly, this doesn't alter the modulus, or "length," of n_1 (see Fig. 10.1). However, the complex square of the coefficient now becomes

$$|e^{i\sigma}n_1 + n_2|^2 = |n_1|^2 + |n_2|^2 + n_1 n_2^* e^{i\sigma} + n_2 n_1^* e^{-i\sigma}$$
$$= |n_1|^2 + |n_2|^2 + 2\Big[\mathrm{Re}(n_1 n_2^*)\cos(\sigma) - \mathrm{Im}(n_1 n_2^*)\sin(\sigma)\Big].$$
$$(10.21)$$

Evidently $|n_1 + n_2|^2 \neq |e^{i\sigma}n_1 + n_2|^2$, even though n_1 and $e^{i\sigma}n_1$ have identical moduli. And indeed, given Fig. 10.2, that's what we would expect.

This process—writing an expansion coefficient as a sum of complex numbers and then modifying one of those numbers by a phase factor—apparently *can* alter probabilities. But does this correspond to anything that actually occurs in quantum mechanics? The answer, we shall see, is Yes.

10.2.3 Unitary Operators Revisited

Unitary operators alter the phases of their eigenstates, and this can alter probabilities. But the process is subtle, and likely eludes many students (and more than a few instructors!).

Discussions of unitary operators are easily found, and they may mention that an overall phase change does not alter a quantum state, while a relative phase change does. Still, explanations of just how and when unitary operators and relative phases can alter probabilities are not standard textbook fare.

Chapter 5 introduced unitary operators and outlined their basic properties. We also saw that unitary operators transform states, and operators, but we didn't discuss their mathematical form, or how, in practice, they act on states.

An arbitrary unitary operator, $\hat{U}$, is defined by the property $\hat{U}^{-1} = \hat{U}^{\dagger}$. This property implies (see Appendix D) that the form of $\hat{U}$ is: $\hat{U} = \exp(i\hat{G}\gamma)$, where γ is a real parameter, and $\hat{G}$ is a Hermitian operator called the *generator* of the transformation effected by $\hat{U}$.

Take the eigenstates of the generator $\hat{G}$ to be the $|\phi_k\rangle$s:

$$\hat{G}|\phi_k\rangle = g_k|\phi_k\rangle. \tag{10.22}$$

This gives $\hat{G}$'s action on the $|\phi_k\rangle$s, but what's $\hat{U}$'s action? In $\hat{U}$, the *operator* $\hat{G}$ appears *in an exponential*. To impart meaning to such an object, the exponential is defined as its Taylor expansion. Thus, $\hat{U}$'s action on the $|\phi_k\rangle$s is

$$\hat{U}|\phi_k\rangle = e^{i\hat{G}\gamma}|\phi_k\rangle = \left\{ 1 + i\hat{G}\gamma + \frac{(i\hat{G}\gamma)^2}{2!} + \frac{(i\hat{G}\gamma)^3}{3!} + \cdots \right\} |\phi_k\rangle$$

$$= \left\{ 1 + ig_k\gamma + \frac{(ig_k\gamma)^2}{2!} + \frac{(ig_k\gamma)^3}{3!} + \cdots \right\} |\phi_k\rangle = e^{ig_k\gamma}|\phi_k\rangle. \tag{10.23}$$

Here I used the fact that $|\phi_k\rangle$ is an eigenstate of $\hat{G}$. Then everything was "collapsed" back down into an exponential. The end result is that an eigenstate of $\hat{G}$ is also an eigenstate of $\hat{U}$, so when $\hat{U}$ acts on an eigenstate of $\hat{G}$, we can simply replace $\hat{G}$ (in $\hat{U}$) with the corresponding eigenvalue. This is a general result: it applies to any unitary operator, its generator, and their eigenstates.

10.2.4 Unitary Operators, Phases, and Probabilities

This section is a rather abstract discussion of the effects of unitary operators on probabilities. If you're content with a concrete example, you can skip to the next section.

If $\hat{U}$ acts on a state $|\Psi\rangle$ that is *not* an eigenstate of $\hat{G}$, we may examine its action by expanding $|\Psi\rangle$ in the eigenstates of $\hat{G}$. Suppose $|\Psi\rangle = \sum\limits_{j=1}^{m} c_j|\phi_j\rangle$; then

$$\hat{U}|\Psi\rangle = e^{i\hat{G}}|\Psi\rangle = \sum_{j=1}^{m} e^{ig_j} c_j|\phi_j\rangle. \tag{10.24}$$

For simplicity, I've set $\gamma = 1$. Evidently the effect of any unitary operator is, at most, the introduction of relative phase factors with respect to the operator's eigenstates—in Eq. (10.24), the e^{ig_j}s are introduced with respect to the $|\phi_k\rangle$s.

Even a *relative* phase change, however, leaves the (complex) square of the coefficient of each $|\phi_j\rangle$ unchanged: $|e^{ig_j}c_j|^2 = |c_j|^2$. And that, in turn, implies (as in Section 10.2.2) that a relative phase change leaves the probabilities unaltered—at least, that is, for observables whose eigenstates are the $|\phi_k\rangle$s.

So how *can* a relative phase change alter probabilities? There's another possibility: introduce a relative phase change with respect to one set of eigenstates, and then examine the probabilities for an observable with *different* eigenstates.

Suppose another orthonormal basis is comprised of the $|\alpha_n\rangle$s, which are *not* eigenstates of $\hat{G}$, but are eigenstates of an operator $\hat{A}$, corresponding to an observable quantity A. Suppose, further, that $\hat{A}$ satisfies

$$\hat{A}|\alpha_n\rangle = \alpha_n|\alpha_n\rangle. \tag{10.25}$$

Then the α_ns, the eigenvalues of A, are the possible measurement values of A.

Take the expansion of the $|\phi_k\rangle$s in the α basis to be

$$|\phi_j\rangle = \sum_{i=1}^{m} a_i^j|\alpha_i\rangle. \tag{10.26}$$

The a_i^js are expansion coefficients. The index j is fixed: it simply denotes the particular $|\phi\rangle$ (in this case, $|\phi_j\rangle$), that we're expanding.

To calculate the probability corresponding to, say, α_β in the state $|\Psi\rangle$, first expand $|\Psi\rangle$ in the α basis:

$$|\Psi\rangle = \sum_{j=1}^{m} c_j|\phi_j\rangle = \sum_{j=1}^{m} c_j \left\{ \sum_{n=1}^{m} a_n^j|\alpha_n\rangle \right\} = \sum_{j,n=1}^{m} c_j a_n^j|\alpha_n\rangle. \tag{10.27}$$

Clearly, the index j now is *not* fixed, reflecting the fact that $|\Psi\rangle$ can include contributions from *all* of the $|\phi_k\rangle$s. From Eq. (10.27), the probability corresponding to α_β is[9]

$$Prob(\alpha_\beta|\Psi) = \left| \sum_{j=1}^{m} c_j a_\beta^j \right|^2 = |c_1 a_\beta^1 + \cdots + c_m a_\beta^m|^2. \tag{10.28}$$

Now let $\hat{U}$ act on $|\Psi\rangle$, denoting the new state $|\Psi_U\rangle$. This produces a relative phase change (see Eq. (10.24)). Then, as in Eq. (10.27), transform

---

[9] If we start out in Eq. (10.27) not with $|\Psi\rangle$, but with $e^{i\theta}|\Psi\rangle$, where $e^{i\theta}$ is an overall phase, we find that the probability corresponding to α_β is unchanged.

to the α basis:

$$|\Psi_U\rangle = \hat{U}|\Psi\rangle = \sum_{j=1}^{m} e^{ig_j} c_j |\phi_j\rangle = \sum_{j=1}^{m} e^{ig_j} c_j \left\{ \sum_{n=1}^{m} a_n^j |\alpha_n\rangle \right\}$$

$$= \sum_{j,n=1}^{m} e^{ig_j} c_j a_n^j |\alpha_n\rangle. \quad (10.29)$$

The probability corresponding to α_β is now

$$Prob(\alpha_\beta|\Psi_U) = \left| \sum_{j=1}^{m} e^{ig_j} c_j a_\beta^j \right|^2 = \left| e^{ig_1} c_1 a_\beta^1 + \cdots + e^{ig_m} c_m a_\beta^m \right|^2. \quad (10.30)$$

The difference between Eqs. (10.30) and (10.28) is analogous to the difference between $|e^{i\sigma} n_1 + n_2|^2$, in Eq. (10.21), and $|n_1 + n_2|^2$, in Eq. (10.20). Evidently, then, the probability in Eq. (10.30) *has* been altered from that in Eq. (10.28).

Note the process that's unfolded. A relative phase change in the ϕ basis didn't alter probabilities with respect to that basis. However, upon transforming to the α basis each $|\alpha_\beta\rangle$'s coefficient was a sum of complex numbers, and the relative phase change in the ϕ basis resulted in an alteration in the phases of these numbers. As we know from Section 10.2.2, this *does* alter probabilities.[10]

We may summarize our results as follows:

> *Statement 1*: The introduction of a relative phase with respect to the eigenstates of an operator $\hat{A}$ can only alter probabilities corresponding to the eigenstates of an operator that does not commute with $\hat{A}$.

Because a unitary operator can, at most, introduce relative phase factors with respect to the operator's eigenstates, we may recast our statement thus:

> *Statement 2*: The maximal effect of any unitary operator $\hat{U}$ is the introduction of relative phases with respect to $\hat{U}$'s eigenstates. This, in turn, can only alter probabilities corresponding to the eigenstates of an operator that does not commute with $\hat{U}$.

10.2.5 Example: A Spin $\frac{1}{2}$ System

Our discussion of unitary operators, relative phases, and probabilities may have seemed a bit abstract. We can make things more concrete with a simple

---

[10] Recall, also, that if two complex numbers are added, the modulus of the sum is altered if a relative phase is introduced between them. See Section 10.1.3 and Fig. 10.2.

example that illustrates how a unitary operator introduces a relative phase change, and how that, in turn, alters probabilities.

In our example we'll again use a spin $\frac{1}{2}$ system. Although it's conventional to take $|+z\rangle$ and $|-z\rangle$ as basis states, many other (really, infinitely many) suitable bases exist. One such basis comprises the states $|+x\rangle$ and $|-x\rangle$. It must, of course, be possible to write $|+z\rangle$ or $|-z\rangle$ in the x basis, and vice versa. It turns out that[11]

$$|\pm x\rangle = \frac{1}{\sqrt{2}}\Big(|+z\rangle \pm |-z\rangle\Big),$$

$$|\pm z\rangle = \frac{1}{\sqrt{2}}\Big(|+x\rangle \pm |-x\rangle\Big). \tag{10.31}$$

Consider a system in the state $|+x\rangle = \frac{1}{\sqrt{2}}(|+z\rangle + |-z\rangle)$, and a unitary operator $\hat{U}_z = \exp(i\hat{S}_z\lambda)$. The probabilities for measurements of S_x in $|+x\rangle$ are $Prob(+\hbar/2) = 1$ and $Prob(-\hbar/2) = 0$.

Now act on $|+x\rangle$ with $\hat{U}_z$, and then transform back to the x basis:

$$
\begin{aligned}
e^{i\hat{S}_z\lambda}|+x\rangle &= \frac{1}{\sqrt{2}}\Big(e^{i\hbar\lambda/2}|+z\rangle + e^{-i\hbar\lambda/2}|-z\rangle\Big) \\
&= \frac{e^{i\hbar\lambda/2}}{2}\Big(|+x\rangle + |-x\rangle\Big) + \frac{e^{-i\hbar\lambda/2}}{2}\Big(|+x\rangle - |-x\rangle\Big) \\
&= \frac{e^{i\hbar\lambda/2}+e^{-i\hbar\lambda/2}}{2}\,|+x\rangle + \frac{e^{i\hbar\lambda/2}-e^{-i\hbar\lambda/2}}{2}\,|-x\rangle \\
&= \cos(\hbar\lambda/2)|+x\rangle + i\sin(\hbar\lambda/2)|-x\rangle. \tag{10.32}
\end{aligned}
$$

Clearly, the probabilities for S_x have changed. They are now: $Prob\,(+\hbar/2) = \cos^2(\hbar\lambda/2)$ and $Prob(-\hbar/2) = \sin^2(\hbar\lambda/2)$.

The operator $\hat{U}_z$ introduced relative phases into the z basis, which, upon transforming back to the x basis, became "mixed." That is, $|+x\rangle$ and $|-x\rangle$ each acquired coefficients consisting of a *sum* (or difference) of the relative phases, $e^{i\hbar\lambda/2}$ and $e^{-i\hbar\lambda/2}$, that were introduced in the z basis.

This example illustrates the minimum condition for a unitary operator to alter probabilities. If $\hat{A}$ represents the observable of interest, then $\hat{A}$ cannot share eigenstates with $\hat{U}$; the system state may be an eigenstate of $\hat{A}$, but not of $\hat{U}$. In the example, $\hat{S}_x$, which represents the observable of interest, does not share eigenstates with $\hat{U}_z$. The system state, $|+x\rangle$, is an eigenstate of $\hat{S}_x$, but not of $\hat{U}_z$.

As pointed out in Section 5.4, unitary operators are of secondary importance for us—with one exception. The time evolution operator plays a central role in quantum mechanics. As we'll see in Chapter 11, Eq. (10.32), with minor modification, is an example of quantum-mechanical time evolution.

---

[11] For example, these relations can be obtained from Townsend (1992), Problem 1.3.

10.3 Wavefunctions

I've ignored wavefunctions so far in this chapter, because they would introduce yet more subtlety into our discussion of the phase. I'll deal with wavefunctions in more depth later, and the phase will afford considerable insight. But since we've now gained some familiarity with the phase, let's at least see what it looks like in a wavefunction, and how our understanding of the phase translates into the language of wavefunctions.

If we restrict consideration to the x axis, then a wavefunction in position representation is a function of x and t, which we'll call $\psi(x,t)$. The wavefunction $\psi(x,t)$ could be transformed into momentum representation. It would then be a function of p and t; call it $\phi(p,t)$. Of course, $\psi(x,t)$ and $\phi(p,t)$ are the *same* quantum state—the *same* wavefunction—written in different representations, or bases.[12]

The position eigenstates are the Dirac deltas, $\delta(x - x_0)$, while the momentum eigenstates are *also* Dirac deltas, but now in momentum space: $\delta(p - p_0)$. Thus, each position eigenstate corresponds to a single, unique position (x value), and each momentum eigenstate corresponds to a single, unique momentum (p value).[13]

Now, how might we think of a function of x, say $f(x)$? If $f(x)$ is a *real* function, then we can think of $f(x)$ as *a collection of real numbers*—one for each x value. Similarly, if $f(x)$ is a *complex* function, as wavefunctions generally are, then $f(x)$ is *a collection of complex numbers*—one for each x value.

If $\psi(x,t)$ is a complex function then, like a complex number, it can be written in polar form.[14] That is, $\psi(x,t)$ may be written in the form $R(x,t)\exp(iS(x,t))$, where the modulus, $R(x,t)$, and the phase, $S(x,t)$, are *real* functions of x and t. Once in polar form, we can "read off" the phase, $S(x,t)$, just as we did for complex numbers in Section 10.1.2.

The continuously distributed complex numbers that constitute a wavefunction are in fact the expansion coefficients. And, just as in the discrete eigenstate cases we considered earlier, multiplication of each such coefficient by the *same* phase constitutes an *overall* phase change.

Now multiply $\psi(x,t)$ by the *overall* phase e^{iA}, where A is a real constant,

$$\psi(x,t) = R(x,t)e^{iS(x,t)} \longrightarrow R(x,t)e^{iA}e^{iS(x,t)}. \tag{10.33}$$

How does this affect the position probability distribution? The original distribution was $|\psi(x,t)|^2 = |R(x,t)e^{iS(x,t)}|^2 = R(x,t)^2$. The

---

[12] Although $\psi(x,t)$ and $\phi(p,t)$ are the same wavefunction—in the sense of the same quantum state—they will not, in general, be of the same functional form.

[13] As pointed out in Section 2.3.2, Dirac deltas are not proper functions.

[14] After all, $\psi(x,t)$ is just a collection of complex numbers.

phase transformation of Eq. (10.33) left this distribution unchanged: $|R(x,t)e^{iA}e^{iS(x,t)}|^2 = |R(x,t)e^{iS(x,t)}|^2$. But this, of course, is what we expect for an overall phase such as $\exp(iA)$.

Similarly, for both the discrete eigenstate case and for wavefunctions, a relative phase change is one that multiplies different coefficients by different phase factors. In the state $\psi(x,t)$, each coefficient is associated with a particular x value. Thus, for the position-space wavefunction $\psi(x,t)$, a *relative phase is a function of* x. For example, multiplying $\psi(x,t)$ by $e^{iX(x)}$,

$$\psi(x,t) = R(x,t)e^{iS(x,t)} \longrightarrow R(x,t)e^{iX(x)}e^{iS(x,t)}, \qquad (10.34)$$

changes $\psi(x,t)$ by a *relative* phase, since $X(x)$ is, in general, different for each different x value.

How does this relative phase change affect the position probability distribution? The original distribution was, again, $R(x,t)^2$. And again, the phase transformation leaves the probability unchanged: $|R(x,t)e^{iX(x)}e^{iS(x,t)}|^2 = R(x,t)^2$. As we now know, this is precisely what a relative phase change *should* do.

It should be clear that exactly analogous arguments apply for the momentum-space wavefunction, $\phi(p,t)$. In that case, of course, the relative phase factor would be a function not of x, but of p, such as $e^{iP(p)}$.

We've seen that a relative phase change can alter probabilities only in a basis other than the one in which the phase change was introduced. The same holds for wavefunctions. The phase change of Eq. (10.34) would lead to an altered momentum probability distribution. Similarly, if a relative phase change was imposed on $\phi(p,t)$, the position probability distribution would be altered. We will revisit these ideas in Chapter 12.

10.4   Reflections

Of what practical value is all this discussion about the complex nature of quantum states, of overall and relative phases, etc.? In one sense, it's *not* of much practical value. Even without understanding such things, if we know how to use the mathematics, we can carry out calculations and obtain predictions. If, for example, we know how to operate with a unitary operator, transform between bases, and extract probabilities, we don't *need* to understand how relative phases lead to altered probabilities.

In another sense, however, understanding the complex nature of quantum states is crucial. Physics, after all, is *not* a rote, turn-the-crank process. It's a *creative* process: we create the physical world—concretely, in the laboratory—and abstractly, in our minds and on paper. To understand the world, to create the world, we must marshal all of our faculties. Without understanding the complex nature of the quantum state, we cannot fully understand how quantum mechanics works—so we cannot bring our full creative power to bear.

10.5 Problems

1. Consider two complex numbers: $2 + 2\sqrt{3}\,i$ and $3 - 3i$.
   (i) Determine the phase of each number.
   (ii) Determine the relative phase, that is, the phase difference, that will:
   (a) maximize the modulus of the sum,
   (b) minimize the modulus of the sum.
   (iii) Sketch the two numbers and their sum on an Argand diagram.

2. Assume that A and B are complex numbers, which may be written in standard form as: $A = \alpha + i\beta$ and $B = \gamma + i\delta$. By manipulating A and B in standard form, show explicitly that the following relations hold.
   (i) $A + A^* = 2i\,\mathrm{Re}A$
   (ii) $A - A^* = 2i\,\mathrm{Im}A$
   (iii) AB is a complex number.
   (iv) $\mathrm{Re}(AB^*) = \mathrm{Re}(A^*B)$
   (v) $(AB)^* = A^*B^*$
   (vi) $|AB|^2 = |A|^2|B|^2$
   (vii) $AB^* + A^*B = 2(\mathrm{Re}A\,\mathrm{Re}B + \mathrm{Im}A\,\mathrm{Im}B)$

3. Now repeat the preceding problem, but working with A and B in polar form, that is: $A = R_A e^{i\theta_A}$ and $B = R_B e^{i\theta_B}$.

4. Write the following as complex numbers in polar form.

   (i) 0   (iii) -1
   (ii) 1   (iv) $i\pi$

5. Draw Argand diagrams that represent the following.
   (i) the function $f(x) = x\exp(ix^2)$ at $x = 2$
   (ii) the function $g(x) = x\{\cos(x) + i\sin(x)\}$
   (iii) the function $h(x) = \sin(x) + ix^2$ at $x = 1$
   (iv) $f(x)$ in the range $1 < x < 2$
   (v) $h(x)$ in the range $0 < x < \frac{\pi}{2}$

6. By considering phases, argue/show whether or not the two states in each of the pairs below are the same state.
   (i) $\frac{1}{\sqrt{2}}\begin{pmatrix} 1 \\ -1 \end{pmatrix}$, $\frac{1}{\sqrt{2}}\begin{pmatrix} 1 \\ 1 \end{pmatrix}$
   (ii) $\frac{1}{\sqrt{2}}\begin{pmatrix} 1 \\ -1 \end{pmatrix}$, $\frac{1}{\sqrt{2}}\begin{pmatrix} -1 \\ 1 \end{pmatrix}$
   (iii) $c_1|1\rangle + c_2|2\rangle$, $c_1 e^{i\theta}|1\rangle + c_2|2\rangle$
   (iv) $c_1|1\rangle + c_2|2\rangle$, $e^{i\theta}\left(c_1|1\rangle + c_2|2\rangle\right)$
   (v) $c_1|1\rangle - ic_2|2\rangle$, $ic_1|1\rangle + c_2|2\rangle$
   (vi) $c_1 e^{i\phi_1}|1\rangle + c_2 e^{i\phi_2}|2\rangle$, $c_1|1\rangle + c_2 e^{i(\phi_2 - \phi_1)}|2\rangle$

7. Explicitly obtain the relations for $|n_1 + n_2|^2$ and for $|e^{i\sigma}n_1 + n_2|^2$ that appear in Section 10.2.1.

(i) Show that if $|\psi\rangle$ is subjected to a phase change of e $|\psi\rangle \rightarrow e^{i\gamma}|\psi\rangle$, where γ is a constant, then there is an *over* change in the τ basis. Show explicitly that this phase chang alter the probabilities associated with the $|\tau\rangle$s.

(ii) Suppose the $|\omega_k\rangle$s (with $k = 1, 2, \ldots, m$) form anothe basis. Show that $|\psi\rangle \rightarrow e^{i\gamma}|\psi\rangle$ cannot alter probabilities in t either.

(iii) Use the foregoing to argue/show that if a quantum state to an overall phase change in *any* basis, then it is subject to overall phase change in *every* basis. Thus, an overall phase any basis implies that *all* probabilities are unchanged. This say that an overall phase change *does not change* a quantu

(iv) As just pointed out, we say that if a quantum state is " such that no probabilities are altered, we have the *same* st argument may seem unsupportable, but similar arguments— whether the *physics* changes—are made in classical physics. Investigate *gauge transformations* in classical electromagneti electricity and magnetism texts discuss such transformatio briefly discuss how such transformations are analogous to a phase change in quantum mechanics.

9. Consider two quantum states $|\psi\rangle$ and $|\psi'\rangle$, where

$$|\psi\rangle = c_1|\phi_1\rangle + c_2|\phi_2\rangle, \qquad |\psi'\rangle = e^{i\delta}c_1|\phi_1\rangle + e^{i\gamma}c_2|\phi_2\rangle.$$

The $|\phi_j\rangle$s satisfy:

$$\hat{A}|\phi_j\rangle = a_j|\phi_j\rangle.$$

(i) Show that the probability of obtaining a_n is the same states (for all n).

(ii) Another set of states, the $|\chi\rangle$s, satisfies:

$$\hat{B}|\chi_k\rangle = b_k|\chi_k\rangle, \quad \text{where}: \ [\hat{A}, \hat{B}] \neq 0.$$

Expanding $|\phi_1\rangle$ and $|\phi_2\rangle$ in the $|\chi\rangle$ basis yields:

$$|\phi_1\rangle = \alpha_1|\chi_1\rangle + \alpha_2|\chi_2\rangle, \qquad |\phi_2\rangle = \beta_1|\chi_1\rangle + \beta_2|\chi_2\rangle.$$

Calculate the probability of obtaining b_1 in a measurement both states, $|\psi\rangle$ and $|\psi'\rangle$. Then calculate and simplify the between the two probabilities.

(iii) Summarize (in words) the results of Parts (i) and (ii).

11
Time Evolution

To-morrow, and to-morrow, and to-morrow,
Creeps in this petty pace from day to day,
To the last syllable of recorded time...
Macbeth[1]

Time—within which we dream our dreams, live through joy and sorrow, and confront the incomprehensibility of our existence—is that by which we know the poignant beauty of life, even as it slips from our grasp.

The time of physics is not invested with the poignancy of "personal time," the time within which we live our lives. Yet even in physics, time plays a special role. Time is the background against which physical processes take place, and the irreversibility of such processes—embodied in the second law of thermodynamics—is fundamentally rooted in the behavior of time, not space.

In much of quantum mechanics, time is irrelevant: the goal is to calculate energy levels by solving the time-*in*dependent Schrödinger equation. Nevertheless, time-dependent quantum systems are fundamentally important: the quantum world is *not* stagnant! This chapter focuses on time-dependent quantum systems, but superpositions, change of basis, Hermitian and unitary operators, complex phase factors, commutators, and uncertainty relations all enter into the discussion. Thus, while time evolution is of great importance in its own right, it will also serve as a means to assemble much of what we've learned so far into a coherent whole.

11.1 The Time-Dependent Schrödinger Equation

Chapter 9 focused on the time-*in*dependent Schrödinger equation. But, as pointed out there, the real starting point is the time-*de*pendent Schrödinger equation:

$$i\hbar\frac{d|\Psi\rangle}{dt} = \hat{H}|\Psi\rangle. \tag{11.1}$$

---

[1] W. Shakespeare, Macbeth, Act 5, scene 5.

Equation (11.1) is a *general* law of quantum mechanics. In particular, no restrictions are placed on the quantum state, $|\Psi\rangle$. It could be a wavefunction, or a spin $\frac{1}{2}$ state, or any other type of quantum state. Furthermore, no representation is specified for $|\Psi\rangle$: it's an *abstract* quantum state, and Eq. (11.1) applies to whatever representation we choose to work in.

Equation (11.1) is, in general, a partial differential equation (since, in practice, $\hat{H}$ typically involves partial derivatives). Partial differential equations are a subject unto themselves and, depending on the form of $\hat{H}$ and the initial state, solving Eq. (11.1) can range from relatively easy to impossibly difficult. Our goal, however, is to attain an understanding of quantum mechanics, not proficiency in applied mathematics. Therefore, we will focus on the most common case, that of time-independent Hamiltonians, that is, $\hat{H} \neq \hat{H}(t)$.

As discussed in Chapter 9, if $\hat{H} \neq \hat{H}(t)$, then Eq. (11.1) is separable into a spatially dependent part and a time-dependent part. Assume we are solving for the jth energy eigenstate (the jth eigenstate of $\hat{H}$), which we'll denote $|E_j\rangle$. Again from Chapter 9, the time dependence of $|E_j\rangle$ is

$$|E_j(t)\rangle = e^{-iE_j t/\hbar}|E_j(0)\rangle, \tag{11.2}$$

that is, the state simply picks up the overall phase $e^{-iE_j t/\hbar}$.

But an overall phase doesn't change a state at all.[2] As such, it may seem that time evolution is dead on arrival—at least for systems with time-independent Hamiltonians. But such is decidedly not the case, as we'll now see. In the process, we'll get an object lesson in the complex nature of the quantum state: the subject matter of Chapter 10.

11.2   How Time Evolution Works

11.2.1   Time Evolving a Quantum State

From Eq. (11.2) we see that an eigenstate of $\hat{H}$, an energy eigenstate, doesn't change in time—it's "stationary." For this reason, eigenstates of $\hat{H}$ are also called *stationary states*. Equation (11.2) gives us the (physically meaningless) time evolution of a stationary state. But how do we time evolve other (i.e. non-stationary) states?

Because $\hat{H}$ is a Hermitian operator—it corresponds to an observable, energy—its (normalized) eigenstates form a complete, orthonormal basis. Thus, we can expand *any* state of a system in terms of its energy eigenstates. Moreover, Eq. (11.2) applies to *any* eigenstate of $\hat{H}$. Thus, Eq. (11.2) provides a prescription for time evolving not only *individual* stationary states, but also *superpositions* of such states—and thus for time evolving any state of the system.

---

[2] Thus, the notation $|E_j(t)\rangle$ or $|E_j(0)\rangle$ is superfluous—just as well to simply write $|E_j\rangle$.

Suppose, for example, that our initial state, denoted $|\chi(0)\rangle$, is a superposition of energy eigenstates: $|\chi(0)\rangle = \sum_k c_k |E_k\rangle$. Then the time-evolved state, $|\chi(t)\rangle$, is simply:

$$|\chi(t)\rangle = \sum_k e^{-iE_k t/\hbar} c_k |E_k\rangle. \tag{11.3}$$

Here we started with $|\chi(0)\rangle$ already expanded in the eigenstates of $\hat{H}$. We can obtain a general equation for $|\chi(t)\rangle$, applicable when $|\chi(0)\rangle$ is *not* so expanded, by working backwards from Eq. (11.3),

$$|\chi(t)\rangle = \sum_k e^{-iE_k t/\hbar} c_k |E_k\rangle = e^{-i\hat{H}t/\hbar} \sum_k c_k |E_k\rangle = e^{-i\hat{H}t/\hbar} |\chi(0)\rangle. \tag{11.4}$$

This distills down to

$$|\chi(t)\rangle = e^{-i\hat{H}t/\hbar} |\chi(0)\rangle = \hat{U}_t |\chi(0)\rangle, \tag{11.5}$$

where I've defined the unitary *time evolution operator*: $\hat{U}_t \equiv e^{-i\hat{H}t/\hbar}$. Equation (11.5) is the general rule for time evolving an arbitrary initial quantum state $|\chi(0)\rangle$, if $\hat{H} \neq \hat{H}(t)$. The simpler Eq. (11.3) still holds once the initial state has been expanded in the eigenstates of $\hat{H}$.

So how, in actual practice, is an arbitrary initial state such as $|\chi(0)\rangle$ time evolved? Basically, Eq. (11.4) is carried out in reverse, that is,

$$|\chi(t)\rangle = e^{-i\hat{H}t/\hbar} |\chi(0)\rangle = e^{-i\hat{H}t/\hbar} \sum_k c_k |E_k\rangle = \sum_k e^{-iE_k t/\hbar} c_k |E_k\rangle. \tag{11.6}$$

Despite this mathematical prescription, students sometimes have difficulty forming a coherent picture of quantum-mechanical time evolution. The following "algorithm" for time evolving a quantum state and extracting measurement probabilities for some observable is designed to provide such a picture.

1. If $\hat{H} = \hat{H}(t)$, solve the time-dependent Schrödinger equation directly.
2. If $\hat{H} \neq \hat{H}(t)$, then,
   (a) Solve the system's time-*in*dependent Schrödinger equation, thus obtaining the eigenstates and eigenvalues of $\hat{H}$.
   (b) Expand the initial state in the eigenstates of $\hat{H}$ (from part (a)).
   (c) Operate on the expanded state with $\hat{U}_t$; substitute the eigenvalues of $\hat{H}$ (from part (a)) as necessary.
3. Solve the eigenvalue equation for the operator corresponding to the observable of interest. The eigenvalues are the possible measured values.
4. Expand the time-evolved state in the eigenstates obtained in step 3.
5. Complex square the expansion coefficients. This yields the probabilities corresponding to the observable's possible measured values.

Mathematically, steps 4 and 5 amount to calculating $|\langle \phi_k | \Psi(t) \rangle|^2$, where $|\Psi(t)\rangle$ is the state of steps 1 or 2, and $|\phi_k\rangle$ is the kth eigenstate obtained in step 3.

This algorithm provides a concise, unified picture of the time evolution of quantum states, and subsequent determination of probabilities. Actually implementing it in any particular case may range from easy to dauntingly difficult.

11.2.2 Unitarity and Phases Revisited

In the example of Section 10.2.5, the unitary operator $\hat{U}_z = \exp(i\hat{S}_z \lambda)$ acted on a spin 1/2 system in the state $|+x\rangle$. By slightly modifying this example we can illustrate quantum-mechanical time evolution concretely.

If a spin $\frac{1}{2}$ system is in a uniform magnetic field in the $+z$ direction, then $\hat{H} = \omega_0 \hat{S}_z$.[3] For our purposes, it's sufficient to set $\omega_0 = 1$. Because $\hat{H} \neq \hat{H}(t)$, the system can be time evolved using $\hat{U}_t = e^{-i\hat{H}t/\hbar} = e^{-i\hat{S}_z t/\hbar}$.

For the case at hand, we already know the solutions to the time-independent Schrödinger equation, and how to expand $|+x\rangle$ in the eigenstates of $\hat{H}$ (see Section 10.2.5). Denoting the time-evolved state $|\psi(t)\rangle$, then, we have

$$|\psi(t)\rangle = e^{-i\hat{S}_z t/\hbar}|+x\rangle \;\; = \;\; \tfrac{1}{\sqrt{2}}\left(e^{-it/2}|+z\rangle + e^{it/2}|-z\rangle\right). \quad (11.7)$$

As in Section 10.2.5, the probabilities for S_x are determined by transforming back to the x basis and collecting the coefficients of $|+x\rangle$ and $|-x\rangle$:

$$|\psi(t)\rangle = \frac{e^{-it/2}}{2}\left(|+x\rangle + |-x\rangle\right) + \frac{e^{it/2}}{2}\left(|+x\rangle - |-x\rangle\right)$$

$$= \left(\frac{e^{-it/2}}{2} + \frac{e^{it/2}}{2}\right)|+x\rangle + \left(\frac{e^{-it/2}}{2} - \frac{e^{it/2}}{2}\right)|-x\rangle$$

$$= \cos(t/2)|+x\rangle - i\sin(t/2)|-x\rangle. \quad (11.8)$$

In the initial state, $|+x\rangle$, the probabilities for S_x were $Prob(+\hbar/2) = 1$ and $Prob(-\hbar/2) = 0$. After time evolution, they are: $Prob(+\hbar/2) = \cos^2(t/2)$ and $Prob(-\hbar/2) = \sin^2(t/2)$.

In the second line of Eq. (11.8), note that both complex numbers comprising the coefficient of $|+x\rangle$ are of modulus $\frac{1}{2}$. But because the numbers' phases vary with t, the modulus of the coefficient itself—the numbers' sum—ranges from 0 to 1. (The coefficient of $|-x\rangle$ exhibits similar behavior.)

All of this constitutes a concrete example not only of time evolution, but also of the concepts developed in Chapter 10. The unitary operator $\hat{U}_t$ introduced relative phases into the initial state (in the z basis). This, of course, did not alter the probabilities of S_z. But upon transforming back to

---

[3] See, for example, Townsend (1992), Section 4.3.

the x basis, $|+x\rangle$ and $|-x\rangle$ each acquired coefficients consisting of a *sum* of complex numbers, with relative phases that $\hat{U}_t$ introduced in the z basis.

11.3 Expectation Values

11.3.1 Time Derivatives

In Chapter 1, we posited three hypothetical worlds. In Worlds 1 and 2—the worlds of classical physics—the basic objects of interest were particles, and the position of a particle was perfectly well defined at all times. World 3 was the world of quantum mechanics. There, a probability function was well defined, so the *average* position of a particle was well defined—but *the* position of a particle was an improper concept (in fact, the particle concept itself was ill defined).

This indistinct nature isn't limited to position—it's a general feature of quantum-mechanical observables. That being the case, *the* value of a quantum observable is, in general, a meaningless concept. But the *average* value of an observable *is* a legitmate concept.[4] Therefore, although it's meaningless to talk of the time dependence of a quantum observable, such as position, it's quite legitimate to talk of the time dependence of its average value, that is, its expectation value.

Recall that the average, or expectation, value of some quantity A, in some state $|\Psi\rangle$, is: $\langle A \rangle = \langle \Psi | \hat{A} | \Psi \rangle$. To calculate the time derivative of $\langle A \rangle$ in the state $|\Psi\rangle$, we treat each "piece" of $\langle \Psi | \hat{A} | \Psi \rangle$ as a distinct mathematical object, and apply the product rule for derivatives:[5]

$$\frac{d\langle A \rangle}{dt} = \frac{d}{dt} \langle \Psi | \hat{A} | \Psi \rangle$$

$$= \left(\frac{d\langle \Psi |}{dt} \right) |\hat{A}|\Psi\rangle + \langle \Psi | \left(\frac{\partial \hat{A}}{\partial t} \right) |\Psi\rangle + \langle \Psi | \hat{A} | \left(\frac{d|\Psi\rangle}{dt} \right)$$

$$= \frac{i}{\hbar} \left(\hat{H} \langle \Psi | \right) \hat{A} | \Psi \rangle + \langle \Psi | \left(\frac{\partial \hat{A}}{\partial t} \right) |\Psi\rangle + \langle \Psi | \hat{A} \frac{-i}{\hbar} \left(\hat{H} | \Psi \rangle \right)$$

$$= \frac{i}{\hbar} \left(\langle \Psi | \hat{H} \hat{A} | \Psi \rangle - \langle \Psi | \hat{A} \hat{H} | \Psi \rangle \right) + \langle \Psi | \left(\frac{\partial \hat{A}}{\partial t} \right) |\Psi\rangle$$

$$= \frac{i}{\hbar} \langle \Psi | [\hat{H}, \hat{A}] | \Psi \rangle + \langle \Psi | \left(\frac{\partial \hat{A}}{\partial t} \right) |\Psi\rangle. \tag{11.9}$$

---

[4] The indistinct nature of observables stems from the inherently probabilistic nature of quantum mechanics. For example, for a (non-Dirac delta) wavefunction, a number obviously can't specify *the* position; but a number *does* specify the *average* value of position.

[5] For $\hat{A} \neq \hat{A}(t)$, the usual case, $|\Psi\rangle$ *must* be the time-evolved state. Otherwise, $\langle A \rangle$ *obviously* can't change!

Note that the Schrödinger equation was used to obtain the third line. We are not so interested in calculational details, however, as in what Eq. (11.9) can tell us about quantum mechanics.

Usually, $\hat{A} \neq \hat{A}(t)$, so that $\partial \hat{A}/\partial t = 0$. Then Eq. (11.9) becomes

$$\frac{d\langle A \rangle}{dt} = \langle \Psi | [\hat{H}, \hat{A}] | \Psi \rangle. \tag{11.10}$$

Consider the case where $|\Psi\rangle$ is an eigenstate of $\hat{H}$ (a stationary state), which we'll call $|E_j\rangle$. Then $\hat{H}|E_j\rangle = E_j|E_j\rangle$. Because $\hat{H}$ is a Hermitian operator, the energy eigenvalue, E_j, must be real. Thus, we have

$$\begin{aligned} \frac{d\langle A \rangle}{dt} &= \langle E_j | \hat{H}\hat{A} | E_j \rangle - \langle E_j | \hat{A}\hat{H} | E_j \rangle \\ &= E_j \left(\langle E_j | \hat{A} | E_j \rangle - \langle E_j | \hat{A} | E_j \rangle \right) = 0. \end{aligned} \tag{11.11}$$

This, of course, is in keeping with what we expect. Apart from a physically meaningless phase, a stationary state does not change in time, so neither can any probabilities—and neither, therefore, can any expectation values.

11.3.2   Constants of the Motion

If $[\hat{H}, \hat{A}] = 0$ (identically), then the observable A is called a *constant of the motion*. This term was presumably carried over from classical mechanics, where we usually focus on actual physical motion. Very often in quantum mechanics, however, we are *not* concerned with physical motion, so the term constant of the motion is slightly unfortunate; a more accurate term would be *constant of the time evolution*.

A constant of the motion is definitely *not* the same thing as a stationary state. It's worth carefully developing the difference beween the two—first, because both are intrinsically important, and second, because doing so further elucidates important concepts.

It's easy to see the significance of a constant of the motion. From Eq. (11.9), if $\hat{A} \neq \hat{A}(t)$, then $[\hat{H}, \hat{A}] = 0$ implies that $d\langle A \rangle/dt = 0$. That is, the expectation value *of the particular observable A* is constant in time *for all states*.

This is distinct from a stationary state—an eigenstate of $\hat{H}$. A stationary state is unaltered by time evolution, so the probability distributions *for all observables* must be unaltered *for this particular state*.

It's interesting to make connection with the discussion of Chapter 10. If $\hat{H} \neq \hat{H}(t)$, then $\hat{U}_t = e^{-i\hat{H}t/\hbar}$. Thus, if $[\hat{H}, \hat{A}] = 0$, then $[\hat{U}_t, \hat{A}] = 0$ also. From Section 10.2.4, this implies that the probability distribution for A cannot be altered under the action of $\hat{U}_t$, regardless of the state. This conclusion—that the *probability distribution* for A is constant—is consistent with, but somewhat stronger than, that obtained in Eq. (11.9), which applies only to the *expectation value* of A.

Moreover, the Schrödinger equation—which was used in obtaining Eq. (11.9)—applies only to time evolution, so Eq. (11.9) cannot be extended to other unitary operators. But because our result from Section 10.2.4 depends only on the behavior of the phase, it holds for any unitary operator. That is, if $[\hat{U}, \hat{A}] = 0$, then the action of $\hat{U}$ cannot alter the probability distribution for the observable A, and this holds not just for time evolution, but for any unitary operator.

11.4 Energy-Time Uncertainty Relations

11.4.1 Conceptual Basis

In Chapter 7 we discussed the general form of the uncertainty relations

$$\Delta A \Delta B \geq \frac{|\langle\, i[\hat{A}, \hat{B}]\, \rangle|}{2}. \tag{11.12}$$

Here A and B are observables, and $\hat{A}$ and $\hat{B}$ are the corresponding Hermitian operators. If we substitute x and p_x for A and B, we obtain

$$\Delta x \Delta p_x \geq \hbar/2, \tag{11.13}$$

the famous uncertainty relation for position and momentum.

The so-called *energy-time uncertainty relation*,

$$\Delta t \Delta E \geq \hbar/2, \tag{11.14}$$

is a useful and important result, and it appears to be quite similar to Eq. (11.12). But there's something curious here: although time appears frequently in quantum mechanics, such as in the time-dependent Schrödinger equation (Eq. (11.1)), a Hermitian operator corresponding to time does *not* appear. You won't see such an operator—because *there is none*! Time is *not* a quantum-mechanical observable, but a parameter. Evidently, then, Eq. (11.12) cannot form the basis for Eq. (11.14). Moreover, the physical interpretation of Eq. (11.12) is quite different from that of Eq. (11.14).

So, where did Eq. (11.14) come from, and what does it mean? To address these questions, first recall how the "standard" uncertainty relations of Eq. (11.12) are interpreted physically. Starting with a system in some particular quantum state, call it $|\Psi\rangle$, a measurement of either A or B is performed. By repeating this process many times—each time starting with the state $|\Psi\rangle$, and each time measuring either A or B—two sets of measurement results are acquired, one set for A, and another for B. The spread in values for the observables A and B—that is, ΔA and ΔB—will be reflected in these measurement results. In other words, ΔA and ΔB characterize, in some sense, the degree to which $|\Psi\rangle$ is a superposition of the eigenstates of $\hat{A}$, and of $\hat{B}$, respectively.

In Eq. (11.14), the physical meaning of ΔE is much the same as the meaning of ΔA and ΔB in Eq. (11.12). That is, repeated measurements of the energy, E, made on a quantum system—each time in the same state—will reflect ΔE. And ΔE (like ΔA and ΔB above) characterizes the degree to which the state is a superposition of energy eigenstates.

The key to interpreting Eq. (11.14) is Δt. Typically, Δt represents a characteristic time for some physical process. Often, it represents some time over which the quantum state changes "appreciably"—an imprecise, though qualitatively accurate, description.

For the spin 1/2 system of Section 11.2.2, Δt may refer to a time interval over which the probabilities of the system could significantly change. For a nucleus or particle that decays, Δt may describe the lifetime of the state.[6] If we're talking about wavefunctions, then Δt may characterize how long it takes a wavepacket to pass some point.

So Δt has no unique meaning, and its interpretation is often not straightforward. Largely because of this, the energy-time uncertainty relation has been a subject of discussion and debate for decades.[7] Still, we have a rough idea of what Δt is, and we also know what it is *not*: Δt is not the spread in a set of "time measurements."

Given the concepts of quantum-mechanical time evolution developed in this chapter, we can at least gain a qualitative understanding of Eq. (11.14). Consider first the limiting case of a stationary state—which, as we've seen, *never* changes. But a stationary state is just an energy eigenstate, and an energy eigenstate clearly exhibits no spread in values for energy measurements, that is, $\Delta E = 0$. In this case, then, $\Delta E = 0$ and $\Delta t \rightarrow \infty$.[8]

If, by contrast, the system state is a superposition of energy eigenstates, then $\Delta E \neq 0$. Roughly speaking, if the system is "more superposed," in the sense of being more "spread out" over the energy eigenstates, then ΔE will be greater.

So, how does ΔE "influence" Δt? Just as there's no unique meaning for Δt, there's no unique answer to this question. Regardless, the qualitative conclusion is the same: the greater the spread in the energy probability distribution—that is, the "more pronounced" the superposition of energy eigenstates—the shorter the time scale over which the system changes appreciably. The situation was well stated by Leslie

---

[6] Quantum mechanics is probabilistic by nature. The lifetime is a probabilistic measure, defined to be that time over which the probability of *not* having decayed is e^{-1}.

[7] See, for example, Ballentine (1998), Section 12.3; Busch (2002).

[8] $\Delta E = 0$ and $\Delta t \rightarrow \infty$ tell us what we want to know about this limiting case; we are unconcerned with strictly (i.e. quantitatively) satisfying Eq. (11.14).

Ballentine: "There is no energy-time relation that is closely analogous to the well-known position-momentum [uncertainty] relation. However, there are several useful inequalities relating some measure of the width of the energy distribution to some aspect of the time dependence."[9] The following example illustrates these ideas quantitatively.

11.4.2 Spin $\frac{1}{2}$: An Example

To make things more quantitative, we can extend the spin 1/2 example of Section 11.2.2. There, $\hat{H} = \hat{S}_z$, the initial state was $|+x\rangle$, and the time-evolved state was,

$$|\psi(t)\rangle = e^{-i\hat{S}_z t/\hbar}|+x\rangle \;=\; \tfrac{1}{\sqrt{2}}\left(e^{-it/2}|+z\rangle + e^{it/2}|-z\rangle\right). \ (11.15)$$

From Eq. (11.15), $Prob\,(E = \hbar/2) = Prob\,(E = -\hbar/2) = 1/2$. Equation (3.10) then yields

$$\Delta E_\psi = \sqrt{\langle E^2\rangle - \langle E\rangle^2} = \sqrt{\tfrac{1}{2}\left(\tfrac{\hbar^2}{4} + \tfrac{\hbar^2}{4}\right) - \left(\tfrac{1}{2}\left[\tfrac{\hbar}{2} - \tfrac{\hbar}{2}\right]\right)^2} = \tfrac{\hbar}{2}\ . \ (11.16)$$

In Section 11.2.2, transforming $|\psi(t)\rangle$ to the x basis led to

$$Prob(S_x = \hbar/2) = \cos^2(t/2), \quad \text{and} \quad Prob(S_x = -\hbar/2) = \sin^2(t/2).$$
$$(11.17)$$

The state of Section 11.2.2 was a maximal superposition of the two energy eigenstates $|+z\rangle$ and $|-z\rangle$, in the sense that it contained equal amounts of each. Let's consider an initial state which is dominated by only one of the energy eigenstates, for example,

$$|\phi(0)\rangle = \tfrac{1}{\sqrt{10}}|+z\rangle + \tfrac{3}{\sqrt{10}}|-z\rangle. \tag{11.18}$$

The time-evolved state is

$$|\phi(t)\rangle = e^{-it/2}\tfrac{1}{\sqrt{10}}|+z\rangle + e^{it/2}\tfrac{3}{\sqrt{10}}|-z\rangle. \tag{11.19}$$

For this state, $Prob(E = \hbar/2) = 1/10$ and $Prob(E = -\hbar/2) = 9/10$. Equation (3.10) then yields $\Delta E_\phi = 3\hbar/10$. Transforming $|\phi(t)\rangle$ to the x basis leads to:

$$Prob(S_x = \pm\hbar/2) = \tfrac{1}{2} \pm \tfrac{3}{10}\cos(t). \tag{11.20}$$

Over what time intervals do $|\psi(t)\rangle$ and $|\phi(t)\rangle$ change appreciably? The answer depends on how we define "appreciable," and that's open to interpretation. To be definite, let's take our definition of an appreciable change

---

[9] Ballentine (1998), p. 347.

to be a change in the probability of some observable of at least, say, .3 (30 percent) from its maximum value.

From Eqs. (11.17) and (11.20), we see that the maximum value of $Prob(S_x = \hbar/2)$ is 1 for $|\psi(t)\rangle$, and 0.8 for $|\phi(t)\rangle$. These maximums occur at $t = 0$ (and also at other times). How much time elapses before these probabilities drop by 0.3 from their maximums? For $|\psi(t)\rangle$, $Prob(S_x = \hbar/2)$ drops to 0.7 at $t \approx 1.16$. For $|\phi(t)\rangle$, $Prob(S_x = \hbar/2)$ drops to 0.5 at $t = \pi/2 \approx 1.57$.[10]

The uncertainties in energy and time for $|\psi(t)\rangle$ and $|\phi(t)\rangle$, then, are

$$\Delta E_\psi = \hbar/2, \qquad \Delta t_\psi \approx 1.16$$
$$\Delta E_\phi = 3\hbar/10, \qquad \Delta t_\phi \approx 1.57 \qquad (11.21)$$

Thus, $\Delta E_\psi > \Delta E_\phi$, and $\Delta t_\psi < \Delta t_\phi$. Qualitatively, this is as expected: greater ΔE implies smaller Δt. Quantitatively, we find that

$$\Delta E_\psi \Delta t_\psi \approx .58\hbar, \qquad \Delta E_\phi \Delta t_\phi \approx .47\hbar . \qquad (11.22)$$

The first relation satisfies Eq. (11.14), while the second violates it only slightly. That's not bad, considering our somewhat arbitrary definition of Δt.

This example provides a case study of the energy-time uncertainty relations, illustrating the interplay between the time scale over which a system changes and how "spread out" it is over its energy eigenstates. Moreover, it puts into practice much of what was discussed in this chapter.

Finally, one topic is conspicuous by its absence from this chapter: the time evolution of wavefunctions. That lacuna will be remedied in the final chapter. There at last, with the fundamental structure of quantum mechanics now at the ready, I will endeavor to build a coherent picture of wavefunctions—including their time evolution.

11.5 Problems

1. Use Eq. (11.9) to show that the expectation value of an observable A does not change with time if the system is in an energy eigenstate, and $\hat{A}$ does not depend explicitly on time. Assume that $\hat{A}$ and $\hat{H}$ do *not* share eigenstates. (You may also assume that $\hat{H} \neq \hat{H}(t)$.)

2. Suppose the initial state of a spin 1 system is

$$|\Psi(0)\rangle = \tfrac{1}{\sqrt{3}}\Big(|1,1\rangle + |1,0\rangle + |1,-1\rangle\Big).$$

Take $\hat{H} = \beta \hat{S}_z^2$, where β is a constant. Find the system's state at some arbitrary later time t. (Carry the calculation as far as possible with the given information.)

---

[10] For both states, similar arguments apply for the case of $S_x = -\hbar/2$.

3. Take the initial state of a spin $\frac{1}{2}$ system to be: $|\Psi(0)\,\rangle = \frac{1}{\sqrt{2}}\Big\{|+z\rangle + |-z\rangle\Big\}$. The system is subject to a Hamiltonian operator that satisfies

$$\hat{H}|\pm z\rangle = E_{\pm}|\pm z\rangle, \qquad \hat{H} \neq \hat{H}(t).$$

(i) Your goal is to determine $Prob(S_y = \hbar/2)$ for the state $|\Psi(t)\rangle$. First outline, step by step, *in words*, how you plan to go about this. Then do it, including carrying out all indicated multiplications.

(ii) Imagine you were *really* doing this—not just as an exercise, but for a research calculation. Then, if you're careful (and if you aren't, you have no business being in physics), you would check that your answer to Part (i) is reasonable. As one such check, state the range of values within which any acceptable probability must lie, and then show that your answer lies within that range for all t.

4. A friend, taking quantum mechanics at another university, tells you the following.

> Our textbook showed that if you start with a free-particle Gaussian and let it time evolve, then a reasonable estimate of the time required for significant spreading is: $T = ma^2/\hbar$. As our professor said, if the mass, m, is $1\,g$, and the width, a, is $.1\,cm$, then the spread time T is roughly $10^{25}s$. Obviously we don't see macroscopic particles spread in the real world, and this explains why.

You could dispute your friend's statement based on (1) the fact that macroscopic objects are comprised of microscopic objects, and (2) the choice of mass is too large. Please provide two *other* clearly stated reasons why the student's argument does *not* adequately explain the non-spreading of macroscopic particles.

5. This problem helps illustrate the connection between quantum time evolution, altering the phases of quantum states, and changing probabilities.

Suppose there exist two independent, orthonormal bases in which to represent some particular quantum state: the alpha basis, comprises $|\alpha_1\rangle$ and $|\alpha_2\rangle$, and the beta basis, comprises $|\beta_1\rangle$ and $|\beta_2\rangle$. Assume, also, that

$$\hat{A}|\alpha_j\rangle = a_j|\alpha_j\rangle, \qquad \hat{B}|\beta_k\rangle = b_k|\beta_k\rangle.$$

(i) Suppose the initial system state is $|\alpha_1\rangle$, and the time evolved the state simply changes by an overall phase factor, $\exp(-i\theta t/\hbar)$ (with θ a constant).

(a) Show explicitly whether or not the probabilities for a_1 and a_2 are changed.

(b) Show explicitly whether or not the probabilities for b_1 and b_2 are changed.

(ii) Now suppose the initial state is $c_1|\alpha_1\rangle + c_2|\alpha_2\rangle$, and the time evolved state is $c_1 \exp(-i\phi_1 t/\hbar)|\alpha_1\rangle + c_2 \exp(-i\phi_2 t/\hbar)|\alpha_2\rangle$, with ϕ_1 and ϕ_2 constants.

(a) Show explicitly whether or not the probabilities for a_1 and a_2 are changed.

(b) Show explicitly whether or not the probabilities for b_1 and b_2 are changed.

6. The Hamiltonian for a spin $\frac{1}{2}$ particle in a uniform magnetic field in the z direction is: $\hat{H} = \omega_0 \hat{S}_z$. At time $t = 0$, the particle is in the state $|\psi(0)\rangle = |+y\rangle$. Note that $|\pm y\rangle = \frac{1}{\sqrt{2}}|+z\rangle \pm \frac{i}{\sqrt{2}}|-z\rangle$.

(i) Find $|\psi(t)\rangle$, the state at time t. Be sure to pull out an overall phase, to simplify your work in Part (ii).

(ii) When is the system first in the state $|-y\rangle$ (after $t = 0$)?

(iii) What is the probability that a measurement of S_y will result in $+\hbar/2$ at time t?

7. The state of a quantum-mechanical system at time $t = 0$ is $|\Psi\rangle$. Suppose that $\hat{G}|\gamma_k\rangle = \gamma_k|\gamma_k\rangle$. Suppose, further, that $|\Psi\rangle = \sum_i c_i|\gamma_i\rangle$.

(i) What is the probability distribution for the possible measurement values associated with $\hat{G}$ in the state $|\Psi\rangle$?

(ii) Suppose the (time-independent) Hamiltonian of the system is $\hat{H} = \alpha \hat{F}^2$ (where α is a constant), and that $\hat{F}$ satisfies: $\hat{F}|\beta_j\rangle = b_j|\beta_j\rangle$. If $|\Psi\rangle = \sum_n d_n|\beta_n\rangle$, find the state of the system at time $t = T$.

(iii) Explain how you would obtain the probability distribution for the possible measurement values associated with $\hat{G}$ at time $t = T$?

8. Sometimes the energy-time uncertainty relation appears in the form:

$$\Delta E \left(\frac{\Delta A}{|d\langle A\rangle/dt|} \right) \geq \frac{\hbar}{2}. \tag{11.23}$$

Suppose that neither $\hat{H}$ nor $\hat{A}$ are explicitly time dependent, and consider the following cases.

(i) The system is in an eigenstate of $\hat{H}$, and $[\hat{H}, \hat{A}] = 0$.

(ii) The system is in an eigenstate of $\hat{H}$, and $[\hat{H}, \hat{A}] \neq 0$.

(iii) The system is not in an eigenstate of $\hat{H}$, and $[\hat{H}, \hat{A}] = 0$.

(iv) The system is not in an eigenstate of either $\hat{H}$ or $\hat{A}$, and $[\hat{H}, \hat{A}] \neq 0$.

(v) The system is not in an eigenstate of $\hat{H}$, but (at the time of each measurement) is in an eigenstate of $\hat{A}$, and $[\hat{H}, \hat{A}] \neq 0$.

Qualitatively (and briefly) describe ΔE, ΔA, and $|d\langle A\rangle/dt|$ for all five cases. Then briefly argue whether each case can be qualitatively reconciled with Eq. (11.23). That is, for each case, can the left-hand side of Eq. (11.23) be greater than zero?

12
Wavefunctions

We thus find that in order to describe the properties of Matter, as well as those of Light, we must employ waves and corpuscles simultaneously. We can no longer imagine the electron as being just a minute corpuscle of electricity: we must associate a wave with it. And this wave is not just a fiction: its length can be measured and its interferences calculated in advance.
Louis de Broglie[1]

1905 was Albert Einstein's *annus mirabilis*: his miracle year. In that one year, young Einstein published five landmark papers. Less appreciated, perhaps, is Erwin Schrödinger's astounding scientific feat in the year 1926. In that one year, Schrödinger published six important papers. Four of these—submitted to *Annalen der Physik* within a span of only six months—together constitute Schrödinger's landmark formulation of quantum mechanics in terms of a wave equation: *wave mechanics*.[2]

As a practical tool, wave mechanics largely supplanted Heisenberg's matrix mechanics. It was wave mechanics that utilized the familiar mathematics of differential equations. And it was wave mechanics that presented at least some kind of picture to aid in one's thinking about quantum physics. Wave mechanics became the tool of choice both for solving many quantum-mechanical problems and for expositions of the subject.

The approach in this book has been different—although wavefunctions have been discussed,[3] they've played a decidedly secondary role. The primary goal has been to build a firm foundation in the fundamentals of quantum mechanics. Now that that foundation is in place, we can use it in a systematic, if brief, exposition of wave mechanics. In so doing, we will uncover the correspondence between the quantum mechanics of wavefunctions and that of discrete eigenstates.

---

[1] Louis de Broglie, Nobel Prize Address, 1929; published in Boorse (1966), p. 1059.

[2] Use of the term "wave mechanics" has gradually declined. The term "quantum mechanics" includes, of course, both wave and matrix mechanics.

[3] See, in particular, Chapters 2, 3, 9, and 10.

12.1   What is a Wavefunction?

12.1.1   Eigenstates and Coefficients

At first glance, it may appear that the quantum mechanics of wavefunctions bears little resemblance to, say, that of a spin $\frac{1}{2}$ system.

A spin $\frac{1}{2}$ state is (at most) a superposition of just two basis states, such as $|+z\rangle$ and $|-z\rangle$; for example:

$$|\alpha\rangle = c_+|+z\rangle + c_-|-z\rangle. \qquad (12.1)$$

A wavefunction, however, is just what its name implies: a continuous function.[4] One possible wavefunction, for example, is a Gaussian centered at x_0: $\Psi(x) = A\exp\left[-B(x-x_0)^2\right]$ (with A and B constants). These two quantum states, $|\alpha\rangle$ and $\Psi(x)$, look *very* different. What connections can we uncover between them?

An obvious discrepancy is that bra-ket notation is used in the spin $\frac{1}{2}$ case, but not for the wavefunction $\Psi(x)$. Bra-ket notation *can* be used for handling wavefunctions, but it's a tricky business—one best deferred until we better understand wavefunctions themselves.

For a spin $\frac{1}{2}$ system, a basis consists of two states—one corresponding to each possible measured value of the observable asociated with a Hermitian operator. For example, $|+z\rangle$ and $|-z\rangle$ correspond to $\hbar/2$ and $-\hbar/2$, respectively, the two possible measured values of S_z (the observable associated with $\hat{S}_z$). Obviously, the basis states are discrete (countable).

Similarly, for a wavefunction there must be a position basis state corresponding to each possible measured value of position, each x (the observable associated with $\hat{x}$). But because the positions, the xs, are continuous, an *infinite* set of position basis states is required to form the position basis.

What about expansion coefficients? For the spin $\frac{1}{2}$ state $|\alpha\rangle$, the coefficients are the complex numbers c_+ and c_-. Their complex squares, $|c_+|^2$ and $|c_-|^2$, are the probabilities of finding the system to be in $|+z\rangle$ or $|-z\rangle$, respectively, upon measurement. If $|\alpha\rangle$ is properly normalized, the sum of these probabilities must equal the total probability, 1.

But where are the expansion coefficients in a wavefunction? The answer is: the set of all expansion coefficients *is* the wavefunction. *Adding* the squares of the expansion coefficients for a system with discrete eigenstates, such as the spin $\frac{1}{2}$ state $|\alpha\rangle$), is analogous to *integrating* over the squares of the expansion coefficients in a state with continuous eigenvalues, such as the wavefunction $\Psi(x)$.[5] If $\Psi(x)$ is properly normalized,

---

[4] Thus, when dealing with wavefunctions, the word *function* is often substituted for *state*.

[5] Consider calculating the total mass in some region of space. If the mass in the region is comprised of discrete objects, we simply sum their individual masses. But if

integrating $|\Psi(x)|^2$ over the entire wavefunction must equal the total probability, 1.

A spin $\frac{1}{2}$ basis state such as $|+z\rangle$ can, of course, be written in *any* spin $\frac{1}{2}$ basis. Written in the z basis (its own eigenbasis), the state $|+z\rangle$ has only one non-zero coefficient. Written in the x basis, however, $|+z\rangle$ is the superposition $\frac{1}{\sqrt{2}}|+x\rangle + \frac{1}{\sqrt{2}}|-x\rangle$; both coefficients are now non-zero.

The position basis states, too, can be written in different representations (we'll see how in Section 12.2.2). If, however, we write a position basis state in the position basis (its own eigenbasis), it's a Dirac delta. For example, $\delta(x-x_0)$ is the position basis state corresponding to x_0, *written in position representation*. This makes sense: if the spin $\frac{1}{2}$ state $|+z\rangle$ is written in its own eigenbasis, there is only one non-zero coefficient. If a position basis state is written in its own eigenbasis, it's a Dirac delta, and is non-zero for only one value of x, that is, only one coefficient is non-zero.

12.1.2 Representations and Operators

Perhaps the term "wavefunction" suggests to you a function of position. Clearly, the wavefunction $\Psi(x)$ *is* a function of position, but that's because we've chosen a specific representation: the position representation. And a wavefunction—like a spin $\frac{1}{2}$ state—can be written in more than one representation. For wavefunctions, there are typically only two representations of interest: the position representation and the momentum representation.

Before considering the momentum representation of a wavefunction, let's briefly return to the spin $\frac{1}{2}$ case. The states $|\pm z\rangle$ are eigenstates of the operator $\hat{S}_z$. They satisfy the eigenvalue equation:

$$\hat{S}_z|\pm z\rangle = \pm\frac{\hbar}{2}|\pm z\rangle, \tag{12.2}$$

again illustrating that if an operator acts on one of its eigenstates, it may be replaced with the corresponding eigenvalue. If an operator acts on a *non*-eigenstate, this no longer holds, but we may expand the state in the operator's eigenstates and *then* act on it. For example, for the state $|\alpha\rangle$ of Eq. (12.1):

$$\hat{S}_z|\alpha\rangle = \hat{S}_z\left(c_+|+z\rangle + c_-|-z\rangle\right) = c_+\frac{\hbar}{2}|+z\rangle - c_-\frac{\hbar}{2}|-z\rangle. \tag{12.3}$$

These are familiar ideas. How do they apply to wavefunctions?

---

the mass is continuously distributed, we must *integrate* the mass *density* (of dimensions mass/length³). Similarly, to obtain the total probability, $|c_+|^2$ and $|c_-|^2$ are simply added, but $|\Psi(x)|^2$ is a probability *density* (of dimensions probability/length), which must be integrated.

For a wavefunction written in position representation, such as $\Psi(x)$, the basis is formed by the position eigenstates. These satisfy

$$\hat{x}|x\rangle = x|x\rangle, \tag{12.4}$$

where $\hat{x}$ is the position operator. Equation (12.4) is analogous to Eq. (12.2). In both equations, an operator acting on its eigenstates is replaced with the corresponding eigenvalue. But, just as Eq. (12.2) does not mean that $\hat{S}_z$, an operator, *equals* $\pm\frac{\hbar}{2}$, Eq. (12.4) does not mean that $\hat{x}$, an operator, *equals* x.

For a wavefunction in position representation, the (linear) momentum operator is: $\hat{p}_x = -i\hbar\frac{\partial}{\partial x}$. As we saw in Chapter 7, $\hat{x}$ and $\hat{p}_x$ do *not* commute, so they do *not* share eigenstates.[6]

What, then, *are* the eigenstates of $\hat{p}_x$? We can answer this by solving a simple differential equation

$$\hat{p}_x f(x) = -i\hbar\frac{df(x)}{dx} = p_x f(x), \tag{12.5}$$

where p_x is the momentum eigenvalue (a number). It's easy to see that Eq. (12.5) is satisfied by $f(x) = N\exp(i\kappa x)$, where N is a normalization constant. Thus, $N\exp(i\kappa x)$ is the eigenstate of $\hat{p}_x$, the linear momentum operator, with eigenvalue $p_x = \hbar\kappa$.

Good enough; but to really understand $f(x)$ demands careful consideration! Let's first describe $f(x)$ in words: $f(x)$ is the eigenstate of momentum corresponding to the eigenvalue p_x, *written in the position representation*. The *position* probability distribution for $f(x)$ is: $|f(x)|^2 = |N\exp(i\kappa x)|^2 = |N|^2$. That is, all positions are *equally probable* for the momentum eigenstate $f(x)$. For a position eigenstate, however, only one, unique x value has a non-zero probability. In this sense, position and momentum eigenstates are as unlike each other as possible.[7]

It might at first seem strange that a state could describe a position probability distribution spread over the entire x axis. But consider a more familiar case.

The state $|+z\rangle$ is an eigenstate of $\hat{S}_z$, which is reflected in the fact that, *in the z basis*, $|+z\rangle$ (obviously) has only a single non-zero term: $|+z\rangle$ itself. And, as an eigenstate of $\hat{S}_z$, $|+z\rangle$ corresponds to a unique S_z value, $\hbar/2$.

But in, say, the x basis, $|+z\rangle$ is a superposition, that is, it's "spread out" over the basis states $|+x\rangle$ and $|-x\rangle$. And, of course, it *must* be spread out: if $|+z\rangle$ could be represented by just *one* x basis state, then there would be only *one* possible measured value for S_x on $|+z\rangle$. But then $|+z\rangle$

---

[6] Note that $\hat{p}_x\Psi(x) = -i\hbar\,d\Psi/dx$ creates a new function, associating a number with each x. Only if each such number depended solely on the corresponding position eigenstate—not, as is the case, on $\Psi(x)$—could the position eigenstates also be eigenstates of $\hat{p}_x$ (compare $\hat{x}$).

[7] Because $f(x)$ extends over the infinite x axis, it's non-normalizable. As discussed in many quantum texts, however, N is often assigned a value for calculational purposes.

would be an eigenstate of $\hat{S}_x$, which it is *not*. Similarly, a momentum eigenstate *must* be spread out in the position representation, otherwise it would be a *position* eigenstate.

12.2 Changing Representations

12.2.1 Change of Basis Revisited

As discussed in Section 10.3, a wavefunction may be expressed in either position or momentum representation, for example, as $\psi(x,t)$ or $\phi(p,t)$. But how does one actually transform between the two representations?

To answer this question, let's revisit non-wavefunction states. Consider a state $|\Gamma\rangle$ that may be expanded in either the $|a_j\rangle$ or $|b_k\rangle$ basis states (where $j, k = 1, 2, \ldots, n$):

$$|\Gamma\rangle = \sum_{j=1}^{n} \langle a_j|\Gamma\rangle |a_j\rangle = \langle a_1|\Gamma\rangle |a_1\rangle + \cdots + \langle a_n|\Gamma\rangle |a_n\rangle, \tag{12.6}$$

$$|\Gamma\rangle = \sum_{k=1}^{n} \langle b_k|\Gamma\rangle |b_k\rangle = \langle b_1|\Gamma\rangle |b_1\rangle + \cdots + \langle b_n|\Gamma\rangle |b_n\rangle. \tag{12.7}$$

Now suppose we wish to transform $|\Gamma\rangle$ from the a basis to the b basis. Conceptually, we know that *each* of the $|b_k\rangle$s will, in general, be composed of contributions from *all* of the $|a_j\rangle$s.[8] Thus, the $|b_1\rangle$ "piece" of $|\Gamma\rangle$ will in general consist of contributions from all n of the $|a_j\rangle$s.

Formally, we could carry out the transformation by acting on $|\Gamma\rangle$, written in the a basis, with the identity operator (the sum of all projection operators), written in the b basis

$$\hat{I}|\Gamma\rangle = \sum_{k=1}^{n} |b_k\rangle\langle b_k|\Gamma\rangle = \sum_{k=1}^{n} |b_k\rangle\langle b_k| \sum_{j=1}^{n} \langle a_j|\Gamma\rangle |a_j\rangle$$

$$= \sum_{j,k=1}^{n} \langle a_j|\Gamma\rangle \langle b_k|a_j\rangle |b_k\rangle. \tag{12.8}$$

By considering only the $k = 1$ term in Eq. (12.8), we again see that the $|b_1\rangle$ piece of $|\Gamma\rangle$ consists, in general, of contributions from all n of the $|a_j\rangle$s.

12.2.2 From x to p and Back Again

How is this relevant to wavefunctions? The last expression in Eq. (12.8) is, in essence, a "recipe" for transforming a state $|\Gamma\rangle$ from the a basis to the b basis. Using this recipe as a guide, we can construct a transformation to take a position-space wavefunction, $\psi(x)$, into momentum space.

---

[8] Similarly, suppose $\hat{x}$, $\hat{y}$, $\hat{z}$, and $\hat{x}'$, $\hat{y}'$, $\hat{z}'$ define two Cartesian coordinate systems. Then the vector $\hat{x}'$, for example, consists in general of contributions from all three unprimed unit vectors.

To see how, let's set $k = 1$ (as suggested above) and $j = 2$ in the last term of Eq. (12.8); this yields: $\langle a_2|\Gamma\rangle\langle b_1|a_2\rangle|b_1\rangle$. The expansion coefficient $\langle a_2|\Gamma\rangle$ is how much $|a_2\rangle$—one of the "old" basis states—there is in $|\Gamma\rangle$. The other inner product, $\langle b_1|a_2\rangle$, is how much $|b_1\rangle$—one of the "new" basis states—there is in $|a_2\rangle$. Therefore, $\langle a_2|\Gamma\rangle\langle b_1|a_2\rangle$ is how much $|b_1\rangle$ is in $|\Gamma\rangle$ due to its $|a_2\rangle$ piece. The total amount of $|b_1\rangle$ in $|\Gamma\rangle$ includes, in general, contributions from all n of the $|a_j\rangle$s—this is why the summation over j appears in Eq. (12.8).

Now construct the transformation of $\psi(x)$ into momentum space. The position eigenstates are now the "old" basis states; the momentum eigenstates, the "new" basis states. In position representation, these states are $\delta(x - x_0)$ and $Ne^{ip_0x/\hbar}$, respectively (with x_0 and p_0 possible position and momentum values).

The expansion coefficient projecting $|\Gamma\rangle$ onto $|a_2\rangle$ was, of course, an inner product. Similarly, the expansion coefficient projecting ψ onto a position basis state, such as $\delta(x - x_0)$, is an inner product; explicitly[9]

$$\langle x_0|\psi\rangle = \int_{-\infty}^{\infty} \delta(x - x_0)\psi(x)dx = \psi(x_0). \qquad (12.9)$$

Corresponding to the inner product $\langle b_1|a_2\rangle$ is the inner product of the position eigenstate, $\delta(x - x_0)$, with the momentum eigenstate,[10] $Ne^{ip_0x/\hbar}$:

$$\langle p_0|x_0\rangle = N\int_{-\infty}^{\infty} e^{-ip_0x/\hbar}\delta(x - x_0)dx = Ne^{-ip_0x_0/\hbar}. \qquad (12.10)$$

In Eq. (12.8), obtaining the total contribution to one of the $|b_k\rangle$s required summing over the old basis states (over j). Similarly, the total contribution to one momentum eigenstate requires integrating over all the old basis states (over dx). Putting all this together, we obtain the following correspondence:[11]

$$\sum_{j=1}^{n}\langle a_j|\Gamma\rangle\langle b_1|a_j\rangle|b_1\rangle \quad\longleftrightarrow\quad N\int_{-\infty}^{\infty} \psi(x)e^{-ip_0x/\hbar}\,dx. \qquad (12.11)$$

We are not yet done, though. Equation (12.11) accounts for the single basis state $|b_1\rangle$ on the left, and the single momentum basis state $Ne^{ip_0x/\hbar}$

---

[9] Note that Eq. (12.9) could be interpreted as explicit confirmation that the numbers comprising a wavefunction are indeed expansion coefficients.

[10] Here the normalization factor, N, is taken to be real.

[11] Nothing corresponding to the ket $|b_1\rangle$ appears on the right-hand side of Eq. (12.11) because a wavefunction, such as $\psi(x)$, is a set of numbers, not numbers and kets. See Section 12.4.1.

on the right. Thus, the integral in Eq. (12.11) does not give us the entire state $\phi(p)$, but only $\phi(p_0)$. To remedy this, we simply replace p_0 with p, that is, we replace the *number* p_0 with the *variable* p (note that this cannot affect the integration). The transformation of $\psi(x)$ into momentum space is, therefore,

$$\phi(p) = N \int_{-\infty}^{\infty} \psi(x) e^{-ipx/\hbar} \, dx. \tag{12.12}$$

We obtained Eq. (12.12) by developing an analogy with Eq. (12.8), so as to illustrate the parallels between the wavefunction case and the discrete eigenvalue case. In fact, however, Eq. (12.12) is a well known and reliable workhorse of applied mathematics: the *Fourier transform*.

As a simple example, let's transform a position eigenstate $\delta(x - x_0)$ into momentum space. From Eq. (12.12) we have:

$$\phi_{x_0}(p) = N \int_{-\infty}^{\infty} \delta(x - x_0) e^{-ipx/\hbar} \, dx = N e^{-ipx_0/\hbar}. \tag{12.13}$$

Evidently the form of a *position* eigenstate in *momentum space*, $Ne^{-ipx_0/\hbar}$, is similar to that of a *momentum* eigenstate in *position space*, $Ne^{ip_0x/\hbar}$. The two differ by a sign change in the exponential, and by the fact that for a position eigenstate, momentum is a variable and position a number (the position eigenvalue), while for a momentum eigenstate, position is a variable and momentum a number (the momentum eigenvalue).

Transforming a momentum eigenstate into momentum space isn't so easy (try it). Nevertheless, it's not hard to see that a momentum eigenstate in momentum space must be a Dirac delta, that is, $\delta(p - p_0)$, in analogy with a position eigenstate in position space: $\delta(x - x_0)$. First, we're dealing with a continuum of eigenvalues, and—just as for position—the Dirac delta is the tool to pick out just one.

Moreover, for momentum eigenstates of the form $\delta(p - p_0)$, an analysis like that above leads to a transformation *from* momentum space *to* position space that's very similar to Eq. (12.12) (another Fourier transform). So if we transform a Dirac-delta momentum eigenstate into position space (much like Eq. (12.13)), we clearly will obtain the proper momentum eigenstate in position space.

12.2.3 Gaussians and Beyond

Analytically evaluating the integral in Eq. (12.12) can range from easy to impossible, depending on the form of $\psi(x)$. An important yet tractable example is that of a Gaussian, such as $\Psi(x) = Ae^{-Bx^2}$ in Section 12.1.1

(taking $x_0 = 0$ for simplicity). For this case, Eq. (12.12) becomes

$$\phi(p) = A \int_{-\infty}^{\infty} e^{-Bx^2} e^{-ipx/\hbar} dx = A \int_{-\infty}^{\infty} e^{-Bx^2 - ipx/\hbar + p^2/4B\hbar^2 - p^2/4B\hbar^2} dx$$

$$= e^{-p^2/4B\hbar^2} \int_{-\infty}^{\infty} e^{-B(x+ip/2B\hbar)^2} dx = \sqrt{\frac{\pi}{B}}\, e^{-p^2/4B\hbar^2}. \quad (12.14)$$

The integration over dx—carried out by completing the square in the argument of the exponential[12]—results in a function of p. Moreover, this momentum-space function is a Gaussian, like its position-space counterpart $\Psi(x)$.

This well-known result—that a Gaussian in position space transforms to another Gaussian in momentum space—displays a deeply pleasing symmetry. That's even more true if we consider the two Gaussians' widths. The width of a Gaussian is conveniently defined as the separation between those points where its value falls to $1/e$ of its maximum. For $\Psi(x) = Ae^{-Bx^2}$, this occurs at $x = \pm B^{-1/2}$, so the width is $2B^{-1/2}$. Similarly, from Eq. (12.14), the width of $\phi(p)$ is $4\hbar B^{1/2}$. Thus, the two Gaussians' widths are inversely proportional.

As the spread in $|\Psi(x)|^2$, the position probability distribution, increases, the spread in $|\phi(p)|^2$, the momentum probability distribution, decreases in inverse proportion, and vice versa. This behavior provides a visualizable, and often used, illustration of the uncertainty relation for position and momentum: $\Delta p \Delta x \geq \frac{\hbar}{2}$ (see Fig. 12.1).

Useful as it is, this illustration can impart an artificially simple picture of wavefunction behavior. Suppose, for example, that we multiply $\Psi(x)$ by

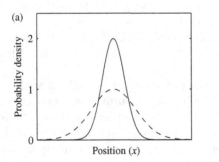
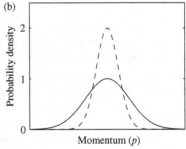

Fig. 12.1   The widths of Gaussian (*a*) position and (*b*) momentum probability distributions are inversely proportional. Solid lines correspond to one state; dashed lines to another.

---

[12] One must also evaluate a deceptively simple *Gaussian integral*, of the form $\int_{-\infty}^{\infty} e^{-\alpha x^2} dx$. These are discussed in many thermal/statistical physics texts, as well as some quantum texts, for example, Townsend (1992).

a phase factor $e^{i\gamma x}$, with γ a real constant. This, of course, doesn't alter the position probability distribution, but if we now calculate the momentum-space wavefunction, as in Eq. (12.14), we obtain a modified $\phi(p)$. For this particular phase factor, $\phi(p)$ will still be a Gaussian (though shifted), but one could imagine countless other phase factors which transform into *non-Gaussian* momentum-space wavefunctions.

There is, in addition, an infinite variety of possible non-Gaussian moduli. The very nice symmetry that exists for a real Gaussian, as embodied in Eq. (12.14), is the exception rather than the rule.

12.3   Phases and Time Evolution

12.3.1   Free Particle Evolution

The time evolution of wavefunctions plays an important role in quantum mechanics. It also provides a means to elucidate the roles of representations and phases in the context of wavefunctions.

Multiplication of a wavefunction $\psi(x, t)$ by an *overall* phase factor e^{iA} (where A is a real constant) results in

$$\psi(x, t) = R(x, t)e^{iS(x,t)} \longrightarrow R(x, t)e^{iA}e^{iS(x,t)}. \qquad (12.15)$$

This leaves the position *and* momentum probability distributions unchanged. By contrast, multiplication by the *relative* phase $e^{iX(x)}$ yields

$$\psi(x, t) = R(x, t)e^{iS(x,t)} \longrightarrow R(x, t)e^{iX(x)}e^{iS(x,t)}, \qquad (12.16)$$

which does *not* alter the position probability distribution, but *does* alter the momentum probability distribution. The situation is analogous for $\phi(p, t)$, the momentum-space wavefunction corresponding to $\psi(x, t)$, in which case a relative phase would be a function of p. All this was discussed in Section 10.3 where, in addition, we saw that this behavior of wavefunctions with respect to phase changes is entirely consistent with what occurs in the discrete eigenstate case.

A free particle provides the simplest example of wavefunction time evolution. The classical Hamiltonian, denoted H_f, is now simply the kinetic energy: $H_f = \frac{p^2}{2m}$. The corresponding Hamiltonian operator is $\hat{H}_f = \frac{\hat{p}^2}{2m}$.

As yet, no representation has been chosen. In position space, $\hat{p} = -i\hbar\frac{d}{dx}$, so that $\hat{H}_f = -\frac{\hbar^2}{2m}\frac{d^2}{dx^2}$. Given this, the position-space time evolution operator is

$$\hat{U}(t) = \exp\left(-\frac{i\hat{H}_f t}{\hbar}\right) = \exp\left(\frac{i\hbar t}{2m}\frac{d^2}{dx^2}\right). \qquad (12.17)$$

An exponential with $\frac{d^2}{dx^2}$ in its argument—what do we do with *this* thing?

We know that the position eigenstates are *not* eigenstates of $\hat{p}$, so they're also not eigenstates of $\hat{H}_f = \frac{\hat{p}^2}{2m}$. And if they're not eigenstates of $\hat{H}_f$, they're not eigenstates of $\hat{U}_t = e^{-i\hat{H}_f t/\hbar}$. If an operator does *not* act on its eigenstates, it can't be replaced by its eigenvalues. So in position space, $\hat{H}_f$ can't be replaced by its eigenvalues.

Clearly, however, the eigenstates of $\hat{p}$ *are* eigenstates of $\hat{H}_f = \frac{\hat{p}^2}{2m}$, and therefore of $\hat{U}_t = e^{-i\hat{H}_f t/\hbar}$. In momentum representation, then, the operator $\hat{p}$ can be replaced with its eigenvalues: $\hat{p} \rightarrow p$. But then $\hat{p}^2$ in $\hat{U}(t)$, also, can be replaced (with p^2).

This suggests time evolving the state in the momentum representation. For an initial state $\phi(p, 0)$, the time-evolved state $\phi(p, t)$ is

$$\phi(p, t) = \hat{U}(t)\phi(p, 0) = \exp\left(-\frac{i\hat{H}_f t}{\hbar}\right)\phi(p, 0)$$

$$= \exp\left(\frac{-i\hat{p}^2 t}{2m\hbar}\right)\phi(p, 0) = \exp\left(\frac{-ip^2 t}{2m\hbar}\right)\phi(p, 0). \quad (12.18)$$

Mathematically, the time evolution of $\phi(p, 0)$ has been reduced to simply multiplying together two functions of p.

Time evolution of a wavefunction, as embodied in Eq. (12.18), is quite analogous to the time evolution of non-wavefunction states, as in Chapter 11. There, an initial state was expanded in the eigenstates of $\hat{U}(t)$, whose action was then simply multiplication of each coefficient by a (time-dependent) phase factor. This is precisely what happens in Eq. (12.18) for a wavefunction.

Note that, in Eq. (12.18), $\phi(p, 0)$ is multiplied by a *relative* phase factor: one that's different for different values of p. And, as expected, this doesn't change the momentum probabilities:

$$\left|\phi(p, 0)\right|^2 = \left|\exp\left(\frac{-ip^2 t}{2m\hbar}\right)\phi(p, 0)\right|^2 = \left|\phi(p, t)\right|^2. \quad (12.19)$$

It does, however, alter the *position* probabilities—again, as expected. To see this, we need to consider how one actually transforms a state from the position representation to the momentum representation and back.

Figure 12.2 (*a*) illustrates the identical position probability distributions for two states (wavefunctions) with identical Gaussian moduli (in position space). One of these states has no phase. The other has a phase factor of the form $\exp\left[i\left\{\frac{x^3}{2} - \frac{3x}{2} - \sin(4x)\right\}\right]$. As a result, the two states' momentum probability distributions vary greatly (see Fig. 12.2 (*b*)). One expects that, because of these different momentum probability distributions, the states will time evolve differently, despite their initially identical position probability distributions.

Again (as in Section 12.2.3), certain dangers are inherent in considering only Gaussians. Consider first a *classical* free particle with initial velocity zero and initial position x_0. The time-evolved system is $x(t) = x_0$: the particle remains at x_0 forever. If any physical system exhibits trivial behavior, surely this is it.

(a)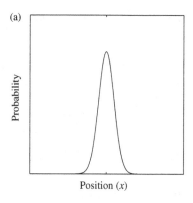

(b)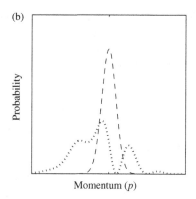

Position (x)   Momentum (p)

Fig. 12.2 (a) Two states with moduli of the form e^{-x^2} (Gaussians) have identical position probability distributions. (b) But if their phases differ, their momentum probability distributions can differ. Here, one has no phase (dashed line), but the other (dotted line) has a phase of the form $e^{i\{x^3/2-3x/2-\sin(4x)\}}$.

Yet the corresponding quantum-mechanical problem is anything but trivial. There is an infinite variety of possible initial states with $\langle x \rangle = x_0$ and $\langle p_x \rangle = 0$. Hardly any of these can be time evolved analytically. The notable exception, once again, is a real Gaussian. It is not hyperbole to say that analytically solving the general free-particle problem is trivially easy in classical mechanics, and impossibly difficult in quantum mechanics.

12.3.2 Wavepackets

Chapter 9 focused largely on energy eigenstates in position space—a particularly important type of wavefunction. But these comprise only a subset of all wavefunctions, and focusing on them can obscure the fact that, in quantum mechanics, particles are typically best represented not as energy eigenstates, but as wavepackets—spatially localized wavefunctions. Section 9.4 considered scattering from step and rectangular potential barriers using energy eigenstates. But the "real" problem involves particles, represented by wavepackets.

One of the more puzzling results of Section 9.4 concerned the step potential for the case $E < V_0$, that is, when the particle energy was less than that of the potential barrier. In that case the probability decreased exponentially inside the barrier, but the physical interpretation of that fact was unclear.

Time evolving a wavepacket incident upon the barrier helps clarify the physics. As for the free particle case, an infinite variety of initial wavepackets is possible. Computer-generated results for an initial Gaussian show that, upon interacting with the barrier, the packet gradually splits up. One piece rapidly forms a reflected packet. But some of the wavefunction penetrates into the barrier, reverses direction, and eventually forms a delayed, reflected packet.

This provides us with a somewhat more picturable situation. Rather than a wave of infinite extent outside the barrier, and exponential decrease within it—all time *in*dependent—the wavepacket exhibits time-*de*pendent behavior that's at least particle-*like*. A wavepacket treatment also lends additional insight into the other cases in Section 9.4, including tunneling.[13]

Many potentials are far more complicated than the step potential. The interaction of particles with such potentials is properly represented, and understood, not through energy eigenstates, but through the time evolution of wavepackets.

12.4 Bra-ket Notation

12.4.1 Quantum States

Because bra-ket notation can be subtle in wave mechanics, I've largely avoided its use in this chapter so far. Now, with the basic concepts of wavefunctions in hand, it's time to examine bra-ket notation in the context of wavefunctions. I'll do so by again comparing the discrete basis state case to wavefunctions.

In either case—discrete basis states or wavefunctions—basis states may be written as kets. For example, suppose that $\{|\alpha_j\rangle\}$, with $j = 1, 2, \ldots, m$, constitutes a set of discrete basis states. A particular ket, such as $|\alpha_3\rangle$, is simply the third basis state in the set.

Corresponding to some position x_k is the position basis state $|x_k\rangle$. But these basis states are continuously distributed along the x axis, so it makes no sense to talk of, say, "the third position ket". And because x is a continuous variable, the ket $|x\rangle$ (without subscript) is taken to represent the complete, infinite set of *all* position kets, all position basis states. The counterpart of $|x\rangle$, then, is not some state $|\alpha_i\rangle$, but the *complete set* $\{|\alpha_j\rangle\}$.

Now suppose that $|\gamma\rangle$ is a state in the space spanned by the set $\{|\alpha_j\rangle\}$. We can take the ket $|\gamma\rangle$ to denote the state *abstractly*, that is, without adopting any specific representation (basis) for the state. Of course, we could choose to expand $|\gamma\rangle$ in, say, the α representation,

$$|\gamma\rangle = \sum_j c_j |\alpha_j\rangle = \sum_j \langle\alpha_j|\gamma\rangle |\alpha_j\rangle. \tag{12.20}$$

The state now appears as a superposition, a weighted sum, of kets.

---

[13] Computer-generated results may be found in, for example, Goldberg (1967); Schiff (1968), pp. 106–107.

By contrast, in matrix mechanics a state appears as a set (*not* a sum) of numbers (*not* kets). For example, $|\gamma\rangle$ would be represented as a column vector, the elements of which are expansion coefficients; if we work in the α basis, these elements are the $\langle\alpha_j|\gamma\rangle$s. The kets themselves do not appear—which is why, to properly interpret a column vector, we must know what basis we're working in.

A wavefunction, also, may be written as an abstract ket, a superposition of basis kets, or a set of numbers (expansion coefficients). This fact is key to developing a correct understanding of Dirac notation as applied to wavefunctions.

Take the *abstract* (representation-free) wavefunction of a system to be ψ, or in bra-ket notation, $|\psi\rangle$. If this state is expanded in position space, it can be written *either* as a superposition of kets *or* as a set of numbers.

In conventional notation, if ψ is expanded in position space and written as a set of numbers, the result is $\psi(x)$, that is, a function of x.[14] In bra-ket notation, this set of numbers appears as $\langle x|\psi\rangle$, so that

$$\psi(x) = \langle x|\psi\rangle. \tag{12.21}$$

Simple though it is, Eq. (12.21) warrants careful interpretation. Really, Eq. (12.21) is more of a translation between conventional and bra-ket notation than an equation in the usual sense. Moreover, "the" inner product $\langle x|\psi\rangle$ is in fact an infinite *set* of inner products, since $|x\rangle$, and thus also $\langle x|$, denotes the complete set of position basis states. That's why what *looks* like a single inner product, a single number, is in fact the entire function $\psi(x)$.

Instead, our wavefunction could be written as a superposition of kets,

$$|\psi\rangle = \int_{-\infty}^{\infty} \langle x|\psi\rangle|x\rangle\,dx. \tag{12.22}$$

Given the replacement of the summation in Eq. (12.20) with an integral in Eq. (12.22), the correspondence between the two equations seems straightforward. Note that both are "dimensionally consistent," in the sense that both sides of each equation consist of kets. Still, interpretation of Eq. (12.22) is quite subtle.

Because $|x\rangle$ denotes the complete set of position basis states, then as just discussed, $\langle x|\psi\rangle$ is the complete set of all such inner products. But if all inner products are already accounted for, and indeed, if $\psi(x)$ already appears in the integrand (as $\langle x|\psi\rangle$), why must we integrate over dx? And what can such an integration even mean?

---

[14] A wavefunction, such as Ae^{-Bx^2}, is effectively a set of numbers (a number is associated with each x value), and thus analogous to a column vector. However, the representation is now obvious: a function of x is in position space; a function of p, in momentum space.

Let's return to Eq. (12.20). Suppose we calculate all the terms—all the weighted kets of the form $\langle \alpha_j | \gamma \rangle | \alpha_j \rangle$—in this equation. This set of weighted kets is *not* the state $|\gamma\rangle$. Only when a superposition is formed—when the terms are added up—do we obtain $|\gamma\rangle$. The integrand in Eq. (12.22), $\langle x | \psi \rangle | x \rangle$, corresponds to the *set* of $\langle \alpha_j | \gamma \rangle | \alpha_j \rangle$s. And just as the $\langle \alpha_j | \gamma \rangle | \alpha_j \rangle$s must be added to obtain $|\gamma\rangle$, so must the $\langle x | \psi \rangle | x \rangle$s be integrated to obtain $|\psi\rangle$.

In physics, the careless use of notation often leads to trouble, and when Dirac notation is applied to wavefunctions, the pitfalls can be treacherous. One might think, for example, that because our wavefunction could be written as either $|\psi\rangle$ or $\langle x | \psi \rangle$, we can write: $|\psi\rangle = \langle x | \psi \rangle$. But this is nonsense (and it even *looks* wrong). Both $|\psi\rangle$ and $\langle x | \psi \rangle$ are legitimate expressions for our wavefunction, but within different frameworks: a *ket* can't equal a set of *numbers*.

One more comment about notation: to denote the same state in different representations, I've used different symbols, such as $\psi(x)$ and $\phi(p)$. This convention makes clear that $\psi(x)$ and $\phi(p)$ are *different functions*, but it obscures the fact that they represent the *same state*.

Instead, the same symbol could be used for both, such as $\psi(x)$ and $\psi(p)$. This convention makes clear that $\psi(x)$ and $\psi(p)$ represent the *same state*, but it obscures the fact that they are *different functions*.

In bra-ket notation, the position and momentum representations of the state $|\psi\rangle$ are $\langle x | \psi \rangle$ and $\langle p | \psi \rangle$, respectively. By contrast, $\langle x | \psi \rangle$ and $\langle p | \phi \rangle$ would normally represent two *different* states, $|\psi\rangle$ and $|\phi\rangle$, written in position and momentum space, respectively. This suggests that the same-symbol convention is more in accord with bra-ket notation than the different-symbol convention. Ultimately, either is acceptable, so long as it's understood what is meant.

12.4.2   Eigenstates and Transformations

Both $|\psi\rangle$ and $\langle x | \psi \rangle$ represent the same quantum state: one as an abstract ket, the other as a function in position space. Similar considerations apply to eigenstates of $\hat{x}$ or $\hat{p}$.

The kets $|x\rangle$ and $|p\rangle$ are the *complete sets* of abstract position and momentum eigenstates, while $|x_0\rangle$, for example, is *one* such state. However, $\langle x | x_0 \rangle$ is the position-space representation of this state: $\langle x | x_0 \rangle = \delta(x - x_0)$. By contrast, $\langle p | x_0 \rangle$ is the state's momentum-space representation: $\langle p | x_0 \rangle = N e^{i p x_0 / \hbar}$.

Because of its inherent importance, and because it provides an opportunity to explore Dirac notation, the Fourier transform (see Section 12.2.2) is worth re-examining. Using $\psi(x) = \langle x | \psi \rangle$ (Eq. (12.21)) and $\langle p | x_0 \rangle = N e^{i p x_0 / \hbar}$, we can construct the transform from position to momentum space in Dirac notation by direct translation from Eq. (12.12):[15]

$$\phi(p) = N \int_{-\infty}^{\infty} \psi(x)e^{-ipx/\hbar}\, dx \longleftrightarrow \int_{-\infty}^{\infty} \langle x|\psi\rangle\langle p|x\rangle dx$$

$$= \int_{-\infty}^{\infty} \langle p|x\rangle\langle x|\psi\rangle dx. \quad (12.23)$$

From Eq. (12.11), we can construct a correspondence between coefficients,

$$\sum_{j=1}^{n}\langle b_1|a_j\rangle\langle a_j|\Gamma\rangle \longleftrightarrow \int_{-\infty}^{\infty}\langle p_0|x\rangle\langle x|\psi\rangle dx. \quad (12.24)$$

The left side is the coefficient of the basis state $|b_1\rangle$, while the right side is the coefficient of the basis state $|p_0\rangle$ (i.e., it's $\phi(p_0)$). Here $\langle a_j|\Gamma\rangle$ corresponds to $\langle x|\psi\rangle$: both are the projections of the respective system states onto the sets of old basis states. In addition, $\langle b_1|a_j\rangle$ corresponds to $\langle p_0|x\rangle$: both are the projections of the sets of old basis states onto one of the new basis states.

The study of wavefunctions constitutes a major part of quantum physics, and we've only scratched the surface. Nothing, for example, has been said about wavefunctions for *non*-free particles: particles subject to forces. But the goal of this chapter, and of this book, has been to develop the basic conceptual and mathematical tools and ideas that underpin the fundamental structure of quantum mechanics. In that, I hope I have achieved some measure of success.

12.5 Epilogue

Our discussion of quantum mechanics now comes to a close. If there is one overarching lesson, it might be this: quantum mechanics is fundamentally about calculating the possible values of physical measurements, and the probabilities of obtaining those values; the quantum state is the entity wherein these probabilities reside. Moreover, while thoughtful people of good will can disagree about the metaphysical meaning of probabilities in quantum mechanics, it's clear how those probabilities are manifested physically: as the statistical results of an ensemble of measurements.

The goal of this book—to develop the tools and ideas necessary to think clearly about quantum mechanics—was relatively modest. Still, you may be

---

[15] Note that $\int_{-\infty}^{\infty}|x\rangle\langle x|dx$ is the identity operator of Section 5.3.2 for position basis states, so the last expression in Eq. (12.23) is really $\langle p|\hat{I}|\psi\rangle$.

left with many questions—both mathematical and conceptual. Take comfort in the fact that you're in good company. As Niels Bohr famously said:[16] "Anyone who is not shocked by quantum theory has not understood it."

In Hinduism, Brahman is the ultimate reality—the great, ineffable spiritual principle of the cosmos. In the Vedas, the ancient textual sources of Hinduism, the indescribability of Brahman is a recurrent theme. So when the disciple seeks to compare Brahman to the things of this world, he is rebuffed: "Not this, not this," he is told. Disconcerting, perhaps—but then why should we expect anything less, anything easily comprehensible, of ultimate reality?

Quantum mechanics is the greatest, the most profound of revolutions in our modern view of the physical world. Even for experts, acheiving a deep conceptual understanding of quantum mechanics can be an elusive goal. Quantum mechanics remains stubbornly recondite, notoriously abstract. Disconcerting, perhaps—but then, as the gateway to a radically new view of physical reality, why should we expect anything less?

Happy thinking!

> The best things can't be told: the second best are misunderstood. Heinrich Zimmer[17]

12.6 Problems

1. Suppose a wave function $\Psi(x)$ is modified by a position-dependent phase, such that: $\Psi(x) \longrightarrow \exp(ip_0x/\hbar)\Psi(x) \equiv \Psi'(x)$, where p_0 is a constant. Show explicitly how, if at all, this will modify $\langle p_x \rangle$, the expectation value for momentum. (It's probably easiest to *not* use bra-ket notation.)

   Sketch some $\phi(p)$, and a corresponding $\phi'(p)$, to illustrate the effect of the above transformation on the momentum amplitude distribution. What physical effect do you think this transformation will have on the state? Discuss your results in the context of phase changes.

2. Suppose $\hat{H}|\beta_k\rangle = E_k|\beta_k\rangle$, where $k = 1, 2, 3$, and that $|\alpha\rangle = \sum_{m=1}^{3} a_m|\beta_m\rangle$. (You can regard $\hat{H}$ as the Hamiltonian, though it's irrelevant for this problem.)

   (i) Write out $\langle \alpha|\hat{H}|\alpha\rangle$ in bra-ket notation; carry the calculation as far as possible with the given information.

   (ii) Clearly and concisely explain why your result is of the proper form for the expectation value of H in the state $|\alpha\rangle$. That is, why does it "make sense" that your equation is that of $\langle H \rangle$?

---

[16] Bohr's comment, though famous, is also mysterious—I have been unable to locate its source.

[17] Quoted in Campbell (1985), p. 21.

(iii) For an arbitrary position-space wavefunction $\psi(x,t)$, the expectation value for position is:

$$\langle\psi|\hat{x}|\psi\rangle = \int_{-\infty}^{\infty} \psi^*(x,t)\ x\ \psi(x,t)dx.$$

Carry out an analysis analogous to that in Part (iii). That is, why should the above integral form an expectation value for position? Identify corresponding mathematical objects in the two expressions. Also, what is the justification for removing the hat from x when it appears inside the integral?

3. The time evolution of a free particle is given by

$$\Psi(x,t) = \int_{-\infty}^{\infty} e^{-ip^2 t/2m\hbar}\langle x|p\rangle\langle p|\psi\rangle dp. \tag{12.25}$$

This equation holds for any initial state. Your job will be to explain how one should interpret the above equation by comparing it to a more familiar case. Assume the equations in the introduction to Problem 2 hold, and that $\hat{H}$ there is a time-independent Hamiltonian.

(i) Write an equation that shows how to time evolve the state $|\alpha\rangle$; carry the calculation as far as possible with the given information.

(ii) Assume the reader is familiar with equations such as the one you wrote in Part (i), but not with Eq. (12.25). By comparing the two equations, explain why Eq. (12.25) gives the time evolved state. Your explanation should focus on comparing/identifying corresponding terms in the two equations; you should also explain why one involves a sum and the other an integral. (If *you* had asked for the explanation, could *you* understand your answer?)

4. Assume that the initial state of a free particle, $\psi(x)$, and the corresponding momentum-space state, $\phi(p)$, are given by:

$$\psi(x) = \frac{1}{\sqrt{\sqrt{\pi}a}}\ e^{-x^2/2a^2}, \qquad \phi(p) = \sqrt{\frac{a}{\hbar\sqrt{\pi}}}\ e^{-p^2 a^2/2\hbar^2}$$

(i) Calculate $\langle x\rangle$ and $\langle x^2\rangle$ for this state.
(ii) Calculate $\langle p\rangle$ and $\langle p^2\rangle$ for this state.
(iii) Show that $\Delta x\Delta p \geq \frac{\hbar}{2}$ is satisfied as an equality for this state.

Note: This problem involves *Gaussian integrals*. Two of them are trivial, *if* symmetries in the integrand and limits are exploited. The other two should be looked up (discussions of Gaussian integrals may be found in many books).

5. Suppose we would like to investigate the time development of the following free-particle initial state:

$$\Psi(x) = Ne^{ip_0 x/\hbar}, \qquad \text{for} \quad a < |x| < (L+a),$$

and $\Psi(x) = 0$, elsewhere. (Take N to be a positive, real normalization constant.) Such a task would ultimately require solution through numerical (computer) methods. Nevertheless, we would want to carry out as much of the calculation as possible using analytical methods. So, do the following.

(i) Sketch $R(x)$, the modulus of $\Psi(x)$.

(ii) Construct the properly normalized initial state (i.e. determine N).

(iii) Construct $\Phi(p)$, the initial momentum-space wavefunction. Simplify your result.

(iv) Write down the explicit form of $\Phi(p, t)$, the time-dependent momentum-space wavefunction, as an integral. Simplify, but don't evaluate, the integral.

Appendix A
Mathematical Concepts

This appendix is intended to provide a very brief review and summary—not a systematic development—of standard mathematical topics used in this book.

A.1 Complex Numbers and Functions

Even basic quantum mechanics demands some familiarity with complex numbers and functions. The starting point for complex analysis is the definition of the number i (also called j): $i \equiv \sqrt{-1}$. From this definition we have: $i^2 = -1$; moreover, $-i^2 = -(ii) = 1$. It's a fascinating fact that a great deal of useful mathematics springs from i, a quantity that, on its face, looks like patent nonsense.

The number i is the simplest example of an *imaginary number*, and multiplication of i by any "ordinary", or real, number, is also an imaginary number. For example, $3i$, $-12i$, and πi are all imaginary numbers.

A *complex number* is comprised of the sum of a real number and an imaginary number, such as $-\pi + 4i$, or $3 - 7i$. Clearly, any complex number may be written as $A + iB$, where A and B are real numbers. Addition, subtraction, multiplication, and division are the same for real and for complex numbers—for complex numbers, of course, we must be careful to properly manipulate i.[1]

By definition, *the complex conjugate* (or just *the conjugate*) of any number, is simply the original number, but with i multiplied by -1 at each occurrence. If C is the complex number $C = A + iB$, then its complex conjugate, denoted C^*, is $C^* = A - iB$. The *complex square* of C, denoted $|C|^2$ (*not* C^2), indicates multiplication of C by C^*: $|C|^2 \equiv C^*C = CC^* = A^2 + B^2$.

A real function, such as $f(x) = \sin(x)$, may be thought of as a collection of real numbers, in the sense that a real number corresponds to each value of the independent variable x. Some functions associate a *complex* number with each value of an independent variable, and so may be thought of as a collection of complex numbers. For example: $g(x) = x^2 + i\cos(x)$ associates a complex number with each x value.

---

[1] In this regard, note that $\pm\frac{1}{i} = \mp i$, which can be seen by multiplying through by i.

The complex conjugate of a function is obtained as for complex numbers: the sign of i is changed everywhere it occurs; thus, $g^*(x) = x^2 - i\cos(x)$. Similarly, the complex square of a function is obtained as it is for a number. For example: $|g(x)|^2 = g(x)g^*(x) = x^4 + \cos^2(x)$. In fact, basic operations on functions go through as for real-valued functions. For example, if $h(x) = e^x - ix^2$, then,

$$g(x)h^*(x) = x^2 e^x + ix^4 + ie^x \cos(x) - x^2 \cos(x)$$
$$= x^2 \left[e^x - \cos(x) \right] + i \left[x^4 + e^x \cos(x) \right]. \quad \text{(A.1)}$$

A crucial difference remains, however, between complex numbers and functions and their real counterparts. An equation involving only real numbers and/or functions is just that: *an* equation. But an equation involving complex quantities may, and often *should*, be thought of as *two* equations. The reason is that each "piece", real and imaginary, must be *independently* satisfied. As a simple example, if α, β, γ, and δ are all real numbers, and if $\alpha + i\beta = \gamma + i\delta$, then we *must* have $\alpha = \gamma$ and $\beta = \delta$.

In this section I've introduced only the bare beginnings of complex analysis. A more in-depth discussion appears in Chapter 10. Still greater depth may be found in almost any mathematical physics text.

A.2 Differentiation

Suppose f is a function of x, that is, $f = f(x)$. The derivative of f with respect to x is written df/dx. Differentiating f (or "taking the derivative" of f) with respect to x yields the rate at which f changes as x is varied. The derivative df/dx is, in general, another function of x. The expression $(df/dx)|_{x=x_0}$ denotes the value of this function when evaluated at $x = x_0$.

Graphically, df/dx is the slope of the graph of $f(x)$ vs. x. Figure A.1 (a) shows such a graph, and the slope at $x = x_0$, that is, $(df/dx)|_{x=x_0}$.

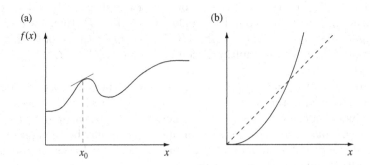

Fig. A.1 (*a*) A function $f(x)$, and its derivative (slope), evaluated at $x = x_0$. (*b*) Sketch of the function $\frac{x^2}{2}$ (solid curve) and its derivative, x (dashed line).

These concepts become more clear if we introduce some specific techniques of differentiation.

Following are some basic derivatives. Take β to be a constant (i.e., $\beta \neq \beta(x)$).

$$\frac{d\beta}{dx} = 0 \qquad \frac{d}{dx}\beta f(x) = \beta \frac{d}{dx} f(x) \qquad \frac{dx^\beta}{dx} = \beta x^{\beta-1}$$

$$\frac{d}{dx}\sin(x) = \cos(x) \qquad \frac{d}{dx}\cos(x) = -\sin(x) \qquad \frac{d}{dx}e^x = e^x \qquad \text{(A.2)}$$

Consider a simple function: $f(x) = x^2/2$. Differentiating, we have: $\frac{d}{dx}\frac{x^2}{2} = x$. These functions are plotted in Fig. A.1(b). Note how these plots "make sense." Inspection of $x^2/2$—the solid curve—reveals that as x increases, so does the *slope* of $x^2/2$ (and the slope is always positive). This behavior is reflected in the plot of the slope—the dashed line in Figure A.1(b). It's also reflected in $df/dx = x$, which (for $x > 0$) is always positive and increases with x.

There are also some important, general rules that apply to differentiation. First, the derivative of a sum is the sum of the derivatives. Thus, for two functions of x, $f(x)$ and $g(x)$, we have

$$\frac{d}{dx}(f+g) = \frac{df}{dx} + \frac{dg}{dx}. \qquad \text{(A.3)}$$

In addition, the derivative of a product of functions, such as $f(x)g(x)$, is

$$\frac{d}{dx}fg = f\frac{dg}{dx} + g\frac{df}{dx}. \qquad \text{(A.4)}$$

In words, the derivative of a product is the first function times the derivative of the second, plus the second times the derivative of the first.

Finally, the "chain rule" is of great importance. Suppose that h is a function of another function, $g(x)$: $h = h(g(x))$. How do we calculate dh/dx? The answer, in essence, is: Take derivatives (i.e. calculate rates of change) for each dependence, until you get to the x dependence. Then multiply these rates together to get the total rate of change. Thus

$$\frac{dh}{dx} = \frac{dh(g(x))}{dg(x)}\frac{dg(x)}{dx}. \qquad \text{(A.5)}$$

For example, if $h = \cos(g(x))$ and $g(x) = x^2$, then

$$\frac{dh}{dx} = \frac{d\cos(x^2)}{dx^2}\frac{dx^2}{dx} = -2x\sin(x^2). \qquad \text{(A.6)}$$

Applications of the chain rule are widespread, and can be quite subtle.

Now suppose a function depends on *two* variables x and y, e.g. $G = G(x,y)$, and suppose we wish to differentiate G with respect to only one of these, say y. To denote such *partial differentiation*, we use a *partial derivative*: $\partial G(x, y)/\partial y$, meaning "take the derivative of G with respect to x, treating y as a constant."

For example, suppose $G = 3xy^2 + \cos(x) + \sin^2(xy)$. Then,

$$\frac{\partial G}{\partial y} = 6xy + 0 + 2\sin(xy)\cos(xy)x. \tag{A.7}$$

Note that differentiation of the last term in G required careful application of the chain rule,

$$\frac{\partial \sin^2(xy)}{\partial y} = \frac{\partial \sin^2(xy)}{\partial \sin(xy)} \frac{\partial \sin(xy)}{\partial xy} \frac{\partial xy}{\partial y} = 2\sin(xy)\cos(xy)x. \tag{A.8}$$

Differentiation—whether ordinary or partial—is usually not terribly difficult. The study of ordinary or partial *differential equations*, however, constitute major and challenging fields of mathematics.

A.3   Integration

The other basic operation in calculus (besides differentiation) is integration. The *indefinite integral* (with respect to x) of a function $\alpha(x)$, written $\int \alpha(x)dx$, is, in general, another function of x. Integrating $\alpha(x)$ (or "taking the integral" of $\alpha(x)$) "adds up" the area under a graph of $\alpha(x)$.

An integral that includes limits is called a *definite integral*. Suppose that $\int \alpha(x)dx = A(x)$; then the correponding definite integral is evaluated thus[2]

$$\int_{x_1}^{x_2} \alpha(x)dx = A(x)\Big|_{x_1}^{x_2} = A(x_2) - A(x_1). \tag{A.9}$$

This integral yields the area under the graph of $\alpha(x)$ between the limits x_1 and x_2. Evaluation of a derivative *at a specific point* yields a number (not a function). So too, evaluation of an integral *over a specific interval* yields a number—the area under the graph within that interval—not a function.

Figure A.2 illustrates integration graphically. The sum of the rectangles' areas approximates the area under the graph. To improve the approximation, we can increase the number of rectangles, and decrease their widths accordingly. If this process is continued until each rectangle is infinitesimally narrow (a mathematical abstraction), we obtain the integral, and the result is exact.

To clarify the meaning of integration, suppose a refinery receives oil both by railcar and pipeline. The volume of, say, the kth railcar, V_k, is known. But for the pipeline, what's known is the time-dependent *rate of*

---

[2] $A(x)$ is not to be confused with the real number A in Section A.1.

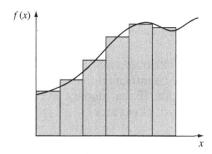

Fig. A.2   The integral of $f(x)$ is the area under the plot of $f(x)$. The sum of the rectangles' areas approximates the exact area (i.e. the integral).

delivery, $R(t)$. Then V_{total}, the total oil delivered to the refinery from time t_i to time t_f, is

$$V_{total} = \sum_j V_j + \int_{t_i}^{t_f} R(t)dt. \tag{A.10}$$

That is, we *sum* the discrete V_ks for all railcars delivered from t_i to t_f, and *integrate* the continuous rate, $R(t)$, from t_i to t_f. In this sense, integration is analogous to summation, but integration applies to continuous distributions.

Finally, it's important to recognize that integration and differentiation are inverses of each other (and for this reason, integrals are sometimes called *anti-derivatives*). For example, if, as above, $\int \alpha(x)dx = A(x)$, then $(d/dx)A(x) = \alpha(x)$. This implies, also, that adding a constant β to the solution of an *in*definite integral is also a solution, because $\frac{d}{dx}A(x) = \frac{d}{dx}[A(x) + \beta]$.

In practice, integration is notoriously challenging, and it's easy to conjure up integrals that are difficult or impossible to solve analytically. Still, integrals often *can* be solved. One simple and general rule is that the integral of a sum is the sum of the integrals.[3] Here, in addition, are a few simple integrals.

$$\int \beta dx = \beta x \qquad \int \beta f(x)dx = \beta \int f(x)dx \qquad \int x^n dx = \frac{x^{n+1}}{n+1}$$

$$\int \sin(x)dx = -\cos(x) \qquad \int \cos(x)dx = \sin(x) \qquad \int e^x dx = e^x$$

$$\tag{A.11}$$

---

[3] The integral of a product, however, is *not* the product of the integrals: $\int f(x)g(x)dx \neq \int f(x)dx \int g(x)dx$. If it were, integration would be immeasurably easier!

Take a few minutes to compare Eqs. (A.11) and (A.2). In particular, note how corresponding derivatives and integrals are indeed inverses of each other.

For more challenging integrals—and they are legion—refer to integration techniques in applied mathematics or mathematical physics texts. Next, try tables of integrals. If that doesn't work, numerical (computer) integration may be required. Good luck.

A.4   Differential Equations

I will not even attempt to present methods of solution of either ordinary or partial differential equations in this section. Rather, our goal is to review what differential equations, and their solutions, *are*.

A differential equation is an equation that involves derivatives of an unknown function. Solving a differential equation, then, is not simply a matter of differentiating a known function (if *that* were the case, solving differential equations would be easy). Rather, it consists of determining the unknown function which satisfies the differential equation.

A simple example is Newton's second law for one-dimensional motion, with constant mass: $F = ma$. This equation describes the trajectory, $x(t)$, of a classical object. It's deceptively simple, for two reasons. First, F may be a function of space and/or time: $F = F(x, t)$. And second, the acceleration, a, is actually the second derivative (in time) of the position, $x(t)$. Thus, with a bit of simple algebra, $F = ma$ becomes

$$\frac{d^2}{dt^2}x(t) - \frac{1}{m}F(x, t) = 0. \tag{A.12}$$

Solving Eq. (A.12)—which requires that $F(x, t)$ be specified—consists of determining the explicit form of the unknown function $x(t)$.

Although *solving* differential equations is often very challenging, *checking* a solution is usually easy. The solution is simply substituted back into the differential equation, derivatives are taken, and if the differential equation is satisfied, the solution is valid.

It's important to realize that, unlike algebraic equations, a differential equation typically has an *infinite* number of valid solutions. To see this, consider the trivially simple case of Newton's second law (Eq. (A.12)) with $F = 0$. Then *any* function of the form $x(t) = a + bt$, with a and b constants, satisfies the differential equation.

Of all valid solutions, however, only one describes a given physical situation. Suppose, for example, that at time $t = 0$ the object described by $x(t) = a + bt$ is at $x = 3$ with speed $v = 2$ in the $+x$ direction. The *particular* solution that describes this physical situation is $x(t) = 3 + 2t$, that is, $a = 3$ and $b = 2$ for this case.

A differential equation that involves only ordinary (non-partial) derivatives, such as Eq. (A.12), is an *ordinary* differential equation, or ODE; if
it involves partial derivatives, it's a *partial* differential equation, or PDE.

The study of PDEs arguably constitutes *the* central subject in applied
mathematics. Rather than superficially discuss some of the many—often
quite sophisticated—methods of attack that exist for PDEs, I'll simply
make a few broad remarks.

Because partial derivatives involve differentiating with respect to one
variable while holding all others constant, PDEs evidently involve multi-
variable functions. For the ordinary differential equation of Eq. (A.12),
the choice of two *numbers*, our inital conditions, selected one particular
solution, corresponding to the physical situation of interest, out of the
infinite set of possible solutions.

A PDE, also, possesses an infinite set of solutions. But selecting out
a particular one of these, corresponding to a particular physical situation,
requires specification not simply of numbers, but of an entire *function*.

The following simple PDE is the traveling wave equation:

$$\frac{\partial^2}{\partial x^2} f(x,t) = v^2 \frac{\partial^2}{\partial t^2} f(x,t). \tag{A.13}$$

Equation (A.13) describes waves on a string, or electromagnetic waves in
vacuum; such waves do not change shape as they travel. Here v is the wave's
speed, and $f(x,t)$ is its amplitude (e.g. for a wave on a string, $f(x,t)$ is the
lateral displacement of the wave at position x and time t).

It can be shown (if one is careful with the chain rule!) that *any* function of the form $f(x-vt)$ satisfies Eq. (A.13).[4] Clearly, there are an infinite
number of such functions. The correct one in any particular case is determined by the initial form of the wave—the "initial condition" is an entire
function.

Another form of wave equation is the time-dependent Schrödinger
equation:

$$-i\hbar \frac{\partial \Psi(x,t)}{\partial t} = \hat{H}\Psi(x,t). \tag{A.14}$$

Equation (A.14) is central to quantum mechanics. In practice, $\hat{H}$ involves
partial derivatives of x, so Eq. (A.14) is in general a highly non-trivial
PDE—in fact, because there's an infinite variety of possible $\hat{H}$s, Eq. (A.14)
represents an infinite variety of PDEs.

Unlike Eq. (A.13), solutions to Eq. (A.14) generally do *not* maintain
their shape as they propagate (see Fig. A.3). Like Eq. (A.13), however,
selection of a particular solution from the infinite set of possibilities requires
specification not simply of a number, but of the initial *function*.

---

[4] To clarify: $f(x,t)$ denotes some as-yet unknown function of x and t; $f(x-vt)$ denotes
any function of x and t with argument $x - vt$, such as $\exp\left[-(x-vt)^2\right]$.

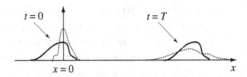

Fig. A.3 Waves satisfying Eq. (A.13) (solid curve) do not change shape from, say, $t = 0$ to $t = T$; those satisfying Eq. (A.14) (dotted curve) typically do.

Again, this discussion of differential equations was intended only to illustrate *what they are*, not to develop methods of solution. For that, consult mathematical physics texts, or treatises devoted to the subject.

Appendix B
Quantum Measurement

The so-called "measurement problem" is probably the single most discussed and debated issue in the foundations of quantum mechanics.[1] Our goal, however, is not to tackle the measurement problem, fascinating and profound though it is, but to understand some practical aspects and implications of quantum measurements, within the framework of the quantum postulates. My focus will be on imparting a clearer conceptual picture of quantum measurement, leaving technical details aside.

In Postulate 2 (of Chapter 2), we introduced the eigenvalue equation $\hat{A}\Psi_k = a_k\Psi_k$. Postulate 3, then, was the following.

> If a measurement of the observable corresponding to the operator $\hat{A}$ is made on the normalized quantum state ψ, given by,
>
> $$\psi = \sum_n c_n\Psi_n, \qquad (B.1)$$
>
> where the Ψ_n's are eigenstates of $\hat{A}$ and the c_n's are expansion coefficients, then a_j, the eigenvalue of Ψ_j, will be obtained with probability $|c_j|^2$. The system will be left in the state Ψ_j immediately after the measurement.

There are three questions we wish to address regarding Postulate 3.

1. How does one actually "make a measurement" in quantum mechanics?
2. Of all possible eigenvalues, what determines which one is obtained in any given measurement?
3. What are the practical implications of leaving the system in the eigenstate corresponding to the eigenvalue that was obtained upon measurement?

A quantum measurement may be thought of as consisting of two distinct but related processes. First, a correlation is established between the

---

[1] A vast literature exists on the measurement problem. Some useful, if rather advanced, references are: Ballentine (1998); and Wheeler (1983). Note that Wheeler (1983) contains (pp. 152–167) a translation of Erwin Schrödinger's paper in which he introduces the famous "Schrödinger's cat". There are, in addition, countless popularizations dealing with the measurement problem.

system to be measured and a measuring device. That is, some physical interaction occurs that couples the state of the micro-system to the state of the macro-system. In particular, the various eigenstates of the observable to be measured are correlated with different states of the measuring device.[2]

These various states of the composite system—eigenstates of the observable and states of the measuring device, coupled together—are themselves coupled to increasingly macroscopic features of the device. At some point, presumably when the interactions approach the macroscopic level, the second process occurs: a particular one of the various possibilities, corresponding to one of the eigenstates, becomes physically manifested in the device. The other possibilities—and the other eigenstates—then simply cease to exist, both in the measuring device and in the quantum system. This process is often called *collapse of the quantum state*.[3] The "argument," in essence, is that this *must* be the case, because we do not observe superpositions of possibilities in the macro-world.

Analysis of the first part of this process, establishing correlations, may be a fearsome mathematical task in any given case.[4] But even if we're not clever enough to figure it out mathematically, there's no problem of principle in imagining some physical interaction that leads to the required correlation between the eigenstates of some observable and measuring device states. (The magnetic field in the Stern–Gerlach device introduced in Chapter 8 served just this role.)

The deep foundational problem stems from the second process. That is, what determines which particular eigenvalue we obtain upon measurement? This largely *is* the quantum measurement problem. There still is no generally agreed upon, satisfactory resolution of this problem (indeed, there's not even universal agreement that it *is* a problem). Quantum mechanics does gives us the probability distribution for the different possible measurement results—but that's *all* it gives us.[5]

Evidently questions 1 and 2 are intertwined. A measurement consists of the establishment of correlations, followed by state collapse. This collapse process, however, offers no explanation of *why* we obtain one eigenvalue rather than another: our second question must remain unanswered.

[2] A discussion of such combined, or "composite," quantum systems is beyond the scope of this book.

[3] This process is also called *collapse of the wavefunction, reduction of the state vector*, etc.

[4] Bohm discusses this correlation process in: Bohm (1951), Chapter 22.

[5] The problem isn't simply *that* collapse occurs, but *how* it does so. Collapse occurs by a process fundamentally different from "normal" quantum time evolution. The question, then, is, What, physically, justifies this process overriding normal time evolution? See in particular Bell (1987), p. 117.

From a calculational standpoint, it is question 3, regarding the practical implications of leaving the system in the eigenstate corresponding to the eigenvalue obtained upon measurement, that is of most interest to us.

The answer is really quite simple. Suppose that at time t_1 our system is in the state ψ of Eq. (B.1). We perform a measurement of $\hat{A}$ and obtain, say, a_3. Then immediately after the measurement the system state is Ψ_3 (see Postulate 3, above). If we want to then time evolve the system from time t_1, *we must start in the state Ψ_3*. This is a straightforward application of Postulate 3, yet it's often overlooked by students.

Similarly, suppose that Ψ_3 is a stationary (time-independent) state (see Chapter 11), and that a second measurement is to be made at some time $t_2 > t_1$. That measurement will be made on the system in Ψ_3, *not* in the state ψ, time evolved until t_2. Again, this is simple, but easily overlooked.

Perhaps part of the difficulty—part of the reason students overlook state collapse—is the way it's manifested in calculations. All terms in a superposition that do *not* correspond to the actual measurement result obtained are simply *erased!* This just *seems* wrong. We generally demand some physical or mathematical justification for "throwing out" part of an equation—but not in this case. The postulate is our only justification, and erasure is precisely what's called for.

Appendix C
The Harmonic Oscillator

The simple harmonic oscillator (SHO) may be the single most important system in physics. The SHO describes vibrations in classical mechanics, electrical oscillations in an LC circuit, quantum-mechanical vibrations of molecules, and countless other physical systems. Even in quantum electro-dynamics—the quantum field theory of the electromagnetic field—the SHO plays a central role, providing a model for photons, the quanta of the field.

The SHO is also one of the few systems for which the time-independent Schrödinger equation (TISE) is exactly soluble. In this appendix, I'll briefly discuss what the SHO is, the SHO solutions to the TISE, and a quite abstract representation of the SHO: the "number representation," which forms the basis for the SHO's entry into quantum field theory.

C.1 Energy Eigenstates and Eigenvalues

Classically, the SHO is a system subject to a "restoring force" that is linear in displacement from the system's equilibrium point. If we set $x = 0$ at that point, then the force is given by: $F = -kx$, where k is a positive constant. (Equivalently, the potential energy is $V = \frac{1}{2}kx^2$.) The usual physical example of a classical SHO is a mass, m, attached to a spring. Such a system oscillates with a characteristic frequency, denoted ω. The parameters m, ω, and k are related by: $\omega = \sqrt{k/m}$.

In quantum mechanics, the SHO can be solved through "standard" means: we form the classical Hamiltonian, and then quantize it, rewriting it in terms of operators. The TISE can be solved exactly for this system, but we'll leave the (non-trivial!) technical aspects aside, focusing instead on results.

The SHO solutions of the TISE are analogous to the energy eigenstates of the infinite potential well (see Chapter 9). That is, they are *energy eigenstates, in position representation*—wavefunctions corresponding to a specific energy.[1]

---

[1] Detailed discussions of the quantum-mechanical SHO appear in nearly every standard quantum mechanics text. Many, such as Griffiths (2005), Liboff (2003), Schiff (1968), and Townsend (1992), include plots of the first few energy eigenfunctions.

For the SHO, the wavefunction boundary conditions are applied at the point where the energy associated with a particular eigenstate equals the oscillator's potential energy. At this point the wavefunction changes from oscillatory to monotonically decreasing—somewhat like what occurs in the step potential and the rectangular barrier (in Chapter 9) for the $E < V_0$ case.

The SHO is a remarkable system in part because of its energy eigenvalues

$$E_n = \hbar\omega \left(n + \tfrac{1}{2}\right), \qquad n = 0, 1, 2, \ldots \tag{C.1}$$

Note first that the eigenvalues are *uniformly spaced*—each separated by $\hbar\omega$ from the next. This is a special property, with special implications.

Let's write the nth SHO energy eigenstate as $|n\rangle$. Suppose we expand an initial SHO state $|\alpha\rangle$ in these eigenstates: $|\alpha\rangle = \sum_j c_j |j\rangle$. Chapter 11 then implies that the time evolved state is

$$|\alpha(t)\rangle = \sum_j e^{-iE_j t/\hbar} c_j |j\rangle = \sum_j e^{-i\omega t \left(j + \frac{1}{2}\right)} c_j |j\rangle$$

$$= \sum_j \left[\cos\left(\omega t \left(j + \tfrac{1}{2}\right)\right) - i \sin\left(\omega t \left(j + \tfrac{1}{2}\right)\right) \right] c_j |j\rangle. \tag{C.2}$$

Equation (C.2) reveals that $|\alpha(t)\rangle$ is periodic, with period $T = 4\pi/\omega$. But *any* SHO state can be expanded in the energy representation, so *any* SHO state is periodic with period T.[2] This is a remarkable property of the SHO, arising from its uniformly spaced energy eigenvalues.

There's another interesting feature of the SHO eigenvalues. Our classical, mass-and-spring SHO could start at $x = 0$ with zero initial velocity. It would then simply remain at rest at $x = 0$; its energy would be precisely zero.

Even for $n = 0$, however, the quantum oscillator's energy is non-zero. This *zero-point energy* is a strictly quantum effect, and it's not unique to the SHO. For example, the energy eigenvalues of the infinite potential well (see Chapter 9) are $E_n = n^2\pi^2\hbar^2/2mL^2$, with $n = 1, 2, 3, \ldots$; the energy can never be zero.

We can at least heuristically justify the zero-point energy by invoking the classical picture of a particle moving along the x axis. Classically, any non-zero energy corresponds to *two* different possible momenta: one in the $+x$ direction, the other in the $-x$ direction. The quantum-mechanical analog to this situation is an energy eigenstate corresponding to two different

---

[2] Actually, T is the longest possible period. If, say, $c_0 = 0$, the period will be less than T.

momenta. This implies an uncertainty in momentum: $\Delta p_x > 0$. This picture fails, however, for a state with zero energy. Then there can be only one momentum, 0, so $\Delta p_x = 0$.[3]

Now consider the zero-point energy in terms of the position-momentum uncertainty relation: $\Delta x \Delta p_x \geq \hbar/2$. For a system constrained to a finite region of the x axis, as are the SHO and the infinite potential well, $0 \leq \Delta x < \infty$. As just argued, a *non*-zero energy eigenstate satisfies $\Delta p_x > 0$, so it's easy to see how the uncertainty relation, also, can be satisfied—that is, how $\Delta x \Delta p_x > 0$. But a *zero*-energy eigenstate satisfies $\Delta p_x = 0$, and since Δx is finite, $\Delta x \Delta p_x > 0$ can't possibly be satisfied.

By this (admittedly non-rigorous) argument, we see that we *must* have a zero-point energy. Interesting as that fact is, however, the zero-point energy takes on a much deeper significance in the context of quantum electrodynamics.

C.2   The Number Operator and its Cousins

As suggested above, the energy eigenstates and eigenvalues of the SHO can be obtained using "standard" methods. An alternative approach uses what are often called operator, or algebraic, methods. The key to such methods is the introduction of two operators, $\hat{a}$ and its adjoint $\hat{a}^\dagger$, defined as:

$$\hat{a} = \sqrt{\frac{m\omega}{2\hbar}} \left(\hat{x} + \frac{i}{m\omega}\hat{p}_x \right), \qquad \hat{a}^\dagger = \sqrt{\frac{m\omega}{2\hbar}} \left(\hat{x} - \frac{i}{m\omega}\hat{p}_x \right). \quad \text{(C.3)}$$

We leave solution of the TISE using $\hat{a}$ and $\hat{a}^\dagger$ to the standard texts. Our concern is the physical meaning of $\hat{a}$, $\hat{a}^\dagger$, and another operator, $\hat{N}$, defined as $\hat{N} \equiv \hat{a}^\dagger \hat{a}$.

Both $\hat{a}$ and $\hat{a}^\dagger$ are *non*-Hermitian operators,[4] and therefore do *not* represent observable quantities (see Postulate 2, in Chapter 2). However, the SHO Hamiltonian operator $\hat{H}$ involves both $\hat{x}$ and $\hat{p}_x$, as do $\hat{a}$ and $\hat{a}^\dagger$. It's not too surprising, then, that $\hat{H}$ can be written in terms of $\hat{a}$ and $\hat{a}^\dagger$; explicitly:

$$\hat{H} = \frac{\hat{p}_x^2}{2m} + \tfrac{1}{2}m\omega^2\hat{x}^2 = \hbar\omega \left(\hat{a}^\dagger\hat{a} + \tfrac{1}{2} \right) = \hbar\omega \left(\hat{N} + \tfrac{1}{2} \right). \quad \text{(C.4)}$$

Evidently the operator $\hat{a}^\dagger\hat{a}$ (or $\hat{N}$) *is* Hermitian, since $\hat{H}$ is Hermitian.[5]
From Equations (C.1) and (C.4), we have

$$\hat{H}|n\rangle = \hbar\omega \left(\hat{N} + \tfrac{1}{2} \right) |n\rangle = E_n|n\rangle = \hbar\omega \left(n + \tfrac{1}{2} \right) |n\rangle, \quad \text{(C.5)}$$

---

[3] Please realize that what's crucial is that the *spread* in p is 0, *not* that $p = 0$ *per se*.
[4] This can be seen by inspection: $\hat{a} \neq \hat{a}^\dagger$, that is, $\hat{a}$ is *not* self-adjoint.
[5] The $\hbar\omega/2$ term simply multiplies a state by a constant, and so cannot affect whether $\hat{H}$ is Hermitian.

where $|n\rangle$ is the nth energy eigenstate. Evidently, the action of $\hat{N}$ on the state $|n\rangle$ is to return the number n: $\hat{N}|n\rangle = n|n\rangle$. For this reason, $\hat{N}$ is called the *number operator.*

To uncover the meaning of $\hat{a}$ and $\hat{a}^\dagger$ themselves, start with the commutation relation,

$$[\hat{a}, \hat{a}^\dagger] = \tfrac{i}{\hbar}(\hat{p}_x\hat{x} - \hat{x}\hat{p}_x) = 1, \tag{C.6}$$

where Eq. (C.3) and the commutation relation $[\hat{x}, \hat{p}_x] = i\hbar$ have been used. Now consider the commutator

$$[\hat{N}, \hat{a}]|\psi\rangle = \left(\hat{a}^\dagger\hat{a}\hat{a} - \hat{a}\hat{a}^\dagger\hat{a}\right)|\psi\rangle = [\hat{a}^\dagger, \hat{a}]\,\hat{a}|\psi\rangle = -\hat{a}|\psi\rangle. \tag{C.7}$$

(Here $|\psi\rangle$ is an arbitrary state, included to clarify the calculation.) Thus, we can write $\hat{N}\hat{a} - \hat{a}\hat{N} = -\hat{a}$, or $\hat{N}\hat{a} = \hat{a}(\hat{N} - 1)$.

Now let's act on an SHO energy eigenstate with $\hat{N}\hat{a}$

$$\hat{N}\hat{a}|n\rangle = \hat{a}(\hat{N} - 1)|n\rangle = \hat{a}\hat{N}|n\rangle - \hat{a}|n\rangle = n\hat{a}|n\rangle - \hat{a}|n\rangle = (n-1)\hat{a}|n\rangle. \tag{C.8}$$

Rewriting Eq. (C.8) as $\hat{N}[\hat{a}|n\rangle] = (n-1)[\hat{a}|n\rangle]$ suggests thinking of $\hat{a}|n\rangle$ as a new state. When the number operator, $\hat{N}$, acts on $\hat{a}|n\rangle$, it returns the number $n-1$. Thus, $\hat{a}|n\rangle = c_-|n-1\rangle$. A similar calculation leads to $\hat{N}[\hat{a}^\dagger|n\rangle] = (n+1)[\hat{a}^\dagger|n\rangle]$, so that $\hat{a}^\dagger|n\rangle = c_+|n+1\rangle$. (Here c_- and c_+ are as-yet undetermined constants.)

Up to the constants c_+ and c_-, then, $\hat{a}^\dagger$ "raises" an SHO energy eigenstate to the next-higher energy eigenstate, and $\hat{a}$ "lowers" an energy eigenstate to the next-lower energy eigenstate. We thus refer to $\hat{a}^\dagger$ and $\hat{a}$ as *raising and lowering operators*, respectively.[6]

Note the similaritites between $\hat{a}$ and $\hat{a}^\dagger$, on the one hand, and $\hat{J}_-$ and $\hat{J}_+$ (of Chapter 8), on the other. All four operators are formed from the sum or difference of Hermitian operators, but the raising and lowering operators themselves are non-Hermitian. And in both cases—angular momentum, or SHO energy—the eigenvalues are uniformly spaced.

Although the raising, lowering, and number operators may seem like curiosities, they can be quite useful in calculations. Moreover, the concepts and methods of the quantum SHO form a cornerstone of quantum electrodynamics, the quantum field theory of electromagnetic interactions.

C.3 Photons as Oscillators

In Newtonian classical mechanics, the fundamental entity is the particle. Only in the late 19th century did fields come to play a central role in

---

[6] By calculating $\langle n|\hat{a}\,\hat{a}^\dagger|n\rangle$ and $\langle n|\hat{a}^\dagger\,\hat{a}|n\rangle$, the inner product of $\hat{a}^\dagger|n\rangle$ with itself and the inner product of $\hat{a}|n\rangle$ with itself, respectively, we find that $c_+ = \sqrt{n+1}$ and $c_- = \sqrt{n}$. Thus, we could create "normalized" raising and lowering operators $\frac{1}{\sqrt{n+1}}\hat{a}^\dagger$ and $\frac{1}{\sqrt{n}}\hat{a}$.

classical physics, in the form of the classical electromagnetic field. At the same time it became clear that light is a manifestation of electromagnetism: an electromagnetic wave.

In quantum mechanics, the fundamental entity is the quantum state—though we typically think of the state as representing particles (though the meaning of "particle" can be unclear in quantum mechanics). Quantum field theory—a subject well beyond the scope of this book—extends quantum physics to describe not just particles, but also fields.[7]

In quantum electrodynamics, the entities that comprise the electromagnetic field—the field "quanta"—are called photons.[8] We need not delve deeply into the concept of a photon; we require only a simple picture of its role in quantum electrodynamics.

For some physical system, let us call each allowable frequency and direction of electromagnetic wave a *mode* of the system. In quantum electrodynamics, the quantum field theory of electromagnetic interactions, each photon in a mode of frequency ω contributes an energy $\hbar\omega$ to that mode. Compare this to a quantum SHO of frequency ω: because its energy eigenvalues are uniformly spaced in steps of $\hbar\omega$, its energy can only be changed in "chunks" of $\hbar\omega$. This suggests pressing the SHO formalism into service to represent photons.

In quantum electrodynamics, photons are routinely created as the electromagnetic field's energy increases, and annihilated as the field's energy decreases. Then the raising, lowering, and number operators take on physical meaning: $\hat{a}^\dagger$ corresponds to creating a photon and $\hat{a}$ to annihilating a photon; as such, $\hat{a}^\dagger$ and $\hat{a}$ are called *creation and annihilation operators*, respectively, in quantum electrodynamics. For a state $|k\rangle$, the number operator, $\hat{a}^\dagger\hat{a}$, again returns k, which now denotes how many photons occupy a particular field mode.

Finally, we return to the SHO's zero-point energy. The similarities between the SHO and photons seem suggestive, but they need not imply that photons actually *are* oscillators. Still, the correspondence is quite deep.

When the electromagnetic field is represented as photons, the SHO's zero-point energy appears as the *vacuum energy*—an energy associated with the field even if there are *no* photons in *any* of the modes.

Although the vacuum energy—the zero-point energy of the field—has very few observable consequences, it's more than a mere curiosity. In quantum electrodynamics, the vacuum energy leads to a picture of the "vacuum" which is not at all empty, but full of activity, including the continual creation and destruction of virtual particle-antiparticle pairs.

---

[7] In quantum field theory, particles are often thought of as localized excitations of fields.

[8] A photon is not simply a "particle of light." See Loudon (2000), pp. 1–2.

To venture further into the vacuum energy, and quantum electrodynamics generally, would lead us into particle physics and quantum field theory proper. These are subjects both wide and deep. If interested, I suggest you consult works that focus on these topics—but be forewarned, it can be very tough going!

Appendix D
Unitary Transformations

In Section 5.4 I stated, without justification, the defining condition for a unitary operator: if $\hat{U}$ is unitary, then $\hat{U}^\dagger = \hat{U}^{-1}$; that is, $\hat{U}$'s adjoint is equal to its inverse. I also pointed out that such operators implement transformations of quantum-mechanical systems in space and time, and that such transformations are connected with conservation principles and symmetries.

In Section 10.2.3 I stated, again without justification, that the general form of a unitary operator is $\hat{U}(\gamma) = \exp(i\hat{G}\gamma)$, with γ a parameter and $\hat{G}$ a Hermitian operator (the time evolution operator, $\exp(-i\hat{H}t/\hbar)$, provides an example).

Although much quantum mechanics can be done without further knowledge of unitary operators and transformations, a deeper acquaintance can be valuable. In this appendix I first discuss the unitarity condition, $\hat{U}^\dagger = \hat{U}^{-1}$, and then introduce infinitesimal unitary transformations. From such a transformation, I obtain its finite counterpart, thus justifying the standard form $\hat{U}(\gamma) = \exp(i\hat{G}\gamma)$. The specific unitary operators of quantum mechanics are then introduced. Finally, I briefly discuss the connection between unitary transformations, symmetries, and conservation principles.

This appendix comprises some of the more difficult material in this book. The hope is that it is nevertheless more condensed and digestible than many of the rigorous, comprehensive treatments found elsewhere.

D.1 Unitary Operators

The laws of physics are preserved under the following transformations in space and time:[1]

- translations (e.g. along the x axis),
- rotations (e.g. about the z axis),
- velocity transformations (i.e., to a uniformly moving frame of reference),[2]
- time translations (time evolution).

---

[1] This statement is discussed in more detail in Section D.3.

[2] Transformations to accelerating reference frames are *not* included, since the laws of physics are *not* preserved under such a transformation.

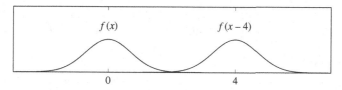

Fig. D.1 $f(x) = e^{-x^2}$ is a Gaussian centered at $x = 0$; $f(x-4) = e^{(x-4)^2}$ is the same Gaussian centered at $x = 4$.

As an example of such a transformation, consider the function $f(x) = e^{-x^2}$, a Gaussian centered at $x = 0$. Then $f(x-4) = e^{(x-4)^2}$ is the same Gaussian translated so that it's centered at $x = 4$ (see Fig. D.1).

This picture of the translation of a function can help us determine the form of operator that would effect such a transformation in quantum mechanics. Suppose that $|\Psi\rangle$ is some normalized quantum state, so that $\langle\Psi|\Psi\rangle = 1$, and also that the operator $\hat{U} = \hat{U}(x)$ translates the state along the x axis, without otherwise altering the state. Then $\hat{U}|\Psi\rangle$ is the translated ket state, and, by definition of the adjoint of an operator (cf. Ch. 5), $\langle\Psi|\hat{U}^\dagger$ is the similarly translated bra state corresponding to $\hat{U}|\Psi\rangle$.

Because we have done nothing other than translate $|\Psi\rangle$ and $\langle\Psi|$, the inner product of these translated states must be unaltered (invariant).[3] That is,

$$\langle\Psi|\Psi\rangle = \left(\langle\Psi|\hat{U}^\dagger\right)\left(\hat{U}|\Psi\rangle\right) = \langle\Psi|\hat{U}^\dagger\hat{U}|\Psi\rangle = \langle\Psi|\left(\hat{U}^\dagger\hat{U}|\Psi\rangle\right). \qquad \text{(D.1)}$$

As indicated by the last expression in Eq. (D.1), if we regard $\hat{U}^\dagger$ as acting to the right, the equality must still hold. But then we must have $\hat{U}^\dagger = \hat{U}^{-1}$. This is precisely the condition that defines a unitary operator.

Analogous arguments could be invoked for the other space-time transformations listed above. We conclude, therefore, that such transformations must be implemented by unitary operators, that is, operators satisfying $\hat{U}^\dagger = \hat{U}^{-1}$.

Because the transformations listed in Section D.1 may be carried out to any extent—for example, a translation may be carried out through any distance—they are called *continuous* transformations.[4]

Consider the special case of translation through an infinitesimal distance. Although this is a mathematical abstraction—we could never *physically* translate anything an infinitesimal distance—it provides the

---

[3] To help see this, consider two real functions, $\alpha(x)$ and $\beta(x)$, and their identically translated counterparts, $\alpha(x-b)$ and $\beta(x-b)$ (with b some constant). The inner product (see Chapter 2) clearly cannot be altered by the translation: $\int_{-\infty}^{\infty} \alpha(x)\beta(x)dx = \int_{-\infty}^{\infty} \alpha(x-b)\beta(x-b)dx$.

[4] An example of a non-continuous, or *discrete*, transformation, is reflection of a function about the origin, for example, $e^{-(x-2)^2} \to e^{-\{(-x)-2\}^2}$.

starting point for finite translations. What sort of operator, denoted $\hat{T}$, will effect a translation through an infinitesimal distance δx? The answer is

$$\hat{T}(\delta x) = \hat{I} + i\hat{G}_T \delta x. \tag{D.2}$$

$\hat{I}$ is the identity operator, so $i\hat{G}_T \delta x$ must effect any change due to the infinitesimal translation. The operator $\hat{G}_T$ is called the *generator* of the translation, and the parameter δx determines *how much* of a translation is carried out—in this case, it's infinitesimal. The i is introduced in anticipation of what's to come (it could be formally eliminated by defining an operator $\hat{G}'_T \equiv i\hat{G}_T$, so that $\hat{T}(\delta x) = \hat{I} + \hat{G}'_T \delta x$).

Now consider again the functions $f(x) = e^{-x^2}$ and $f(x-4) = e^{-(x-4)^2}$. An inverse transformation takes us back to the original function. To be explicit, define $g(x) \equiv f(x-4)$; that is, $g(x) = e^{-(x-4)^2}$. If we now translate $g(x)$ by 4 in the $-x$ direction, we have $g(x+4) = e^{-\{(x+4)-4\}^2} = e^{-x^2}$. That is, we recover our original function. In fact, for all of the space-time transformations listed above, we expect that the original function can be recovered by an inverse transformation, similar to translating $f(x)$ in the $+x$ direction and then back again.

This must hold for our infinitesimal translation, also. That is, we must have: $\hat{T}^{-1}(\delta x)\hat{T}(\delta x) = \hat{I}$. This, in turn, will hold if $\hat{T}^{-1}(\delta x) = \hat{I} - i\hat{G}_T \delta x$:[5]

$$\hat{T}^{-1}(\delta x)\hat{T}(\delta x) = \left(\hat{I} - i\hat{G}_T \delta x\right)\left(\hat{I} + i\hat{G}_T \delta x\right)$$

$$= \hat{I}^2 - i\delta x \hat{G}_T \hat{I} + i\delta x \hat{I} \hat{G}_T = \hat{I} + i\delta x \left(\hat{G}_T - \hat{G}_T\right) = \hat{I}. \tag{D.3}$$

Now, because $\hat{T}$ is unitary, we must have $\hat{T}^{-1} = \hat{T}^\dagger$, that is,

$$\hat{I} - i\hat{G}_T \delta x = (\hat{I} + i\hat{G}_T \delta x)^\dagger. \tag{D.4}$$

Note that Eq. (D.4) has *not* been derived; rather, it is a condition that must be *enforced*. How are we to do so?

Consider a ket $|\psi\rangle = c_1|a_1\rangle + c_2|a_2\rangle$, the corresponding bra $\langle\psi| = c_1^*\langle a_1| + c_2^*\langle a_2|$, and an operator $i\hat{A}$ that satisfies $i\hat{A}|a_k\rangle = ia_k|a_k\rangle$. Then, defining a ket $|\psi'\rangle \equiv i\hat{A}|\psi\rangle$, we have: $|\psi'\rangle = ia_1c_1|a_1\rangle + ia_2c_2|a_2\rangle$. From Section 4.2.1, the bra $\langle\psi'|$ corresponding to the ket $|\psi'\rangle$ must be:

$$\langle\psi'| = i^*a_1^*c_1^*\langle a_1| + i^*a_2^*c_2^*\langle a_2| = -ia_1^*c_1^*\langle a_1| - ia_2^*c_2^*\langle a_2|. \tag{D.5}$$

Thus, the operator that, acting to the left on $\langle\psi|$, yields $\langle\psi'|$, is $-i\hat{A}^\dagger$. That is, $(i\hat{A})^\dagger = -i\hat{A}^\dagger$. The result of this argument is that Eq. (D.4) is indeed satisfied, i.e., $\hat{T}$ is unitary, *if* we insist that $\hat{G}_T = \hat{G}_T^\dagger$; that is, if $\hat{G}_T$ is Hermitian.

---

[5] In Eq. (D.3), factors of $\hat{I}$ are dropped. The term $\hat{G}_T\hat{G}_T(\delta x)^2$ is also dropped, because it is of second order in the infinitesimal quantity δx.

D.2 Finite Transformations and Generators

If we carry out the infinitesimal translation through δx repeatedly—in fact, approaching an infinite number of times—we obtain a translation through some *finite* distance x.

Formally, we can rewrite δx as:

$$\delta x = \lim_{N \to \infty} \frac{x}{N}. \tag{D.6}$$

A finite translation of some function $f(x)$ is then of the form:

$$\hat{T}(x)f(x) = \ \dots \ \left(\hat{I} + i\hat{G}_T \frac{x}{N}\right)\left(\hat{I} + i\hat{G}_T \frac{x}{N}\right)\left(\hat{I} + i\hat{G}_T \frac{x}{N}\right)f(x), \tag{D.7}$$

where I've shown only the first three of the $N \to \infty$ infinitesimal transformations. From Eq. (D.7), then,[6]

$$\hat{T}(x) = \lim_{N \to \infty} \left(\hat{I} + i\hat{G}_T \frac{x}{N}\right)^N = e^{ix\hat{G}_T}. \tag{D.8}$$

Unitary operators—including those that effect the space-time transformations of Section D.1—are of the form of $\hat{T}(x)$: an exponential whose argument includes i, a parameter (such as x), and a Hermitian generator (such as $\hat{G}_T$).

So far, we've said that $\hat{G}_T$ is the generator of translations, but what *specific* operator is $\hat{G}_T$? We can largely answer this question independent of quantum mechanics. Consider translating some function $F(x)$ in the $+x$ direction by a small increment Δx (see Fig. D.2). The slope of $F(x)$ at x_0

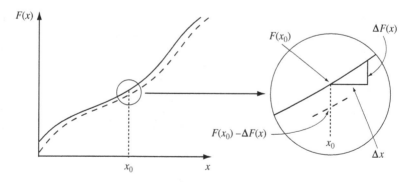

Fig. D.2 Translation of a function $F(x)$ in the $+x$ direction by a small increment Δx. Here $\Delta F(x) \approx \Delta x \frac{dF}{dx}\big|_{x_0}$. In the limit $\Delta x \to \delta x$, this approximation becomes exact.

---

[6] To verify Eq. (D.8), multiply out the first few terms in the limit and collect powers of $i\hat{G}_T x/N$. Then compare this with the Taylor expansion of the exponential.

is $\frac{dF}{dx}\big|_{x_0}$. In general, the slope is different at each x value. Within the small interval Δx, however, it's approximately constant, so for the value of the translated function at x_0 we can write,

$$F(x_0 - \Delta x) = F(x_0) - \Delta F(x) \approx F(x_0) - \Delta x \frac{dF}{dx}\bigg|_{x_0}$$

$$= F(x_0) - run\frac{rise}{run} = F(x_0) - rise. \quad \text{(D.9)}$$

Here the terms "rise" and "run" have been introduced to aid understanding.

In the limit $\Delta x \to \delta x$, $F(x)$ is translated only *infinitesimally*, and Eq. (D.9) becomes exact:

$$F(x_0 - \delta x) = F(x_0) - \delta x \frac{dF}{dx}\bigg|_{x_0}. \quad \text{(D.10)}$$

Transcribing Eq. (D.10) into an infinitesimal translation *operator*, we have

$$\hat{T}(\delta x) = \hat{I} - \delta x \frac{d}{dx}. \quad \text{(D.11)}$$

Equation (D.11) makes no reference to quantum mechanics. For quantum wavefunctions in position space, the linear momentum operator is: $\hat{p}_x = -i\hbar\frac{d}{dx}$. Thus, in quantum mechanics, $\hat{G}_T = -\hat{p}_x/\hbar$, and $\hat{T}(\delta x)$ becomes

$$\hat{T}(\delta x) = \hat{I} - \frac{i(\delta x)\hat{p}_x}{\hbar}. \quad \text{(D.12)}$$

Equations (D.6) and (D.8) then imply that

$$\hat{T}(x) = e^{-i\hat{p}_x x/\hbar}, \quad \text{(D.13)}$$

which is indeed the correct finite quantum-mechanical translation operator.

Similar arguments lead to the unitary operators for boosts and rotations (although the latter case is a bit subtle). The unitary operator for time translations is obtained from the time-dependent Schrödinger equation (see Chapter 11). The resulting unitary operators are

$$\hat{T}(x) = e^{-i\hat{p}_x x/\hbar},$$

$$\hat{R}_z(\theta) = e^{-i\hat{J}_z \theta/\hbar},$$

$$\hat{V}(v) = e^{i\hat{x}p_x/\hbar},$$

$$\hat{U}(t) = e^{-i\hat{H}t/\hbar}, \quad \text{(D.14)}$$

which correspond to translation by x, rotation about the z axis by θ, a velocity boost of v, and time translation by t.[7]

---

[7] In quantum mechanics, the operators themselves, exclusive of $\hbar$, are called the generators. Thus, $\hat{p}_x$, not $-\hat{p}_x/\hbar$, is the generator of translations, and similarly for $\hat{x}$, $\hat{H}$, and $\hat{J}_z$, the z angular momentum operator.

D.3 Continuous Symmetries

D.3.1 Symmetry Transformations

This is all quite elegant, but what is its physical significance? Although the crucially important time evolution operator was introduced, that's also done in Chapter 11, and without recourse to such sophisticated methods. The physical significance of the preceding lies largely in the connection of unitary transformations to conserved quantities.

This area of quantum mechanics can be quite difficult to grasp, so a careful exposition of terminology is warranted. In particular, what is meant by the terms *symmetry* and *continuous symmetry transformation*?

If a cylinder is rotated about its axis through *any* angle, it looks the same after the rotation—its appearance is unchanged. Rotated off-axis, however, the cylinder looks different. We say that the cylinder's appearance is *invariant* (unchanged), or symmetric, with respect to rotation about its axis, but *not* with respect to off-axis rotations.[8]

Now consider translating the cylinder along its axis. Apart from this shift, the cylinder looks the same. This illustrates the importance of terminology: We cannot say whether the cylinder is invariant under translation along its axis until we choose a definition of invariance for this particular case.

Symmetries in physics are often more abstract than that of a cylinder rotated about its axis, but they are united by the fact that when we say a physical system exhibits a symmetry, we mean that *some feature of the system is invariant under some transformation*. This seems vague, but it must be to encompass the many and varied transformations, and corresponding invariant quantities, in physics.

A *symmetry transformation*, then, is simply some transformation under which a symmetry is exhibited, that is, under which some feature of the system is invariant. A *continuous* symmetry transformation is one for which the symmetry is exhibited for *all* values of some continuously varying parameter. Rotation of a cylinder about its axis (through *any* rotation angle) is one example.

D.3.2 Symmetries of Physical Law

Continuous symmetries in physics can be classified in a number of ways: geometrical or dynamical, internal or external, local or global.[9] For our purposes, however, I only wish to distinguish those symmetries that are inherent in Nature herself from those that are associated with a particular physical system.

---

[8] Even off-axis rotations leave the cylinder's appearance invariant for certain rotation angles, such as 2π. But these constitute *discrete*, rather than continuous, symmetries.

[9] These classifications, and other relevant aspects of continuous symmetries, are discussed in a lucid yet non-technical way in: Kosso (2000).

As pointed out at the beginning of Section D.1, the laws of physics are (so far as we know) invariant with respect to translations, rotations, velocity transformations, and time translations. This statement deserves careful consideration: it does *not* mean that the numerical values of physical quantities, such as position or velocity, are preserved, but that physical *laws*, such as Newton's laws of motion, or the Schrödinger equation, remain valid. Moreover, "invariance" is used here in a somewhat different sense than before. A symmetry of a physical system refers to the invariant nature of some feature of the system (e.g., the appearance of a cylinder). In contrast, the laws of physics are not a feature of any particular system, but of Nature herself.

Invariance of the laws of *classical* physics is easily illustrated. A passenger on a train throws a ball directly upwards, and sees it travel straight up, and straight back down. To a trackside observer, however, the ball does not describe a simple linear path, but (because of the train's motion) an arc. Yet in both cases Newton's laws—in particular the second law, $\vec{F}_{net} = \frac{d\vec{p}}{dt}$—are obeyed.

How is the invariance of physical law reflected in quantum mechanics? One way is through the invariance of inner products, enforced by the unitarity condition, $\hat{U}^\dagger = \hat{U}^{-1}$, developed in Section D.1:

$$\langle\phi|\hat{U}^\dagger\hat{U}|\theta\rangle = \langle\phi|\hat{U}^{-1}\hat{U}|\theta\rangle = \langle\phi|\theta\rangle. \tag{D.15}$$

Physically, this invariance of inner products reflects the fact that the projection of one state onto another, and thus all probabilities, cannot depend on whether we perform an experiment in a laboratory "here", or in a lab that is translated (in space *or* time), rotated, or uniformly moving with respect to it.

Yet even classically, this isn't how the world *seems* to work. For example, compare dropping a ball near Earth's surface to dropping it a few thousand kilometers above the surface. Then the difference in observed behavior of the ball does not simply arise from an altered point of view. The ball actually behaves differently, because of the altered gravitational field (although in both cases, the laws of classical physics are still obeyed).

The quantum-mechanical situation is similar. For example, if an atom is in an electric field, it possesses more energy levels than when it is not. Thus, translating an atom into or out of such a field actually changes its spectrum.

The problem is again one of terminology. The invariance of physical law under translation refers to translation of the entire physical situation—fields and all—in space. Thus, our reference to the invariance of experimental results due to translation or rotation of the laboratory implicitly assumes that everything upon which the experiment's outcome depends is contained within the laboratory itself.

The invariance of physical law provides the foundation for our use of unitary operators. I now turn to symmetries of particular systems.

D.3.3 System Symmetries

Because of the invariance of physical laws discussed in Section D.3.2, *all* translations, rotations, boosts, and time translations are continuous symmetry transformations. But symmetries may also be associated with a particular feature of a particular system, and in that context, only those transformations for which that system feature is invariant are properly called symmetry transformations.

Our cylinder's appearance was invariant for on-axis rotations, but not for off-axis rotations. In contrast, a sphere is invariant with respect to rotations about any axis that passes through its center. Many other shapes aren't invariant with respect to *any* rotations (pick up a stone and rotate it). In this context, whether a rotation is a continuous symmetry transformation depends on the particular "system" (i.e. the particular shape).

In quantum mechanics, it's $\hat{H}$ that characterizes a particular system, so it's $\hat{H}$'s properties that characterize a system's symmetries. Recall (from Section 5.4) that an operator $\hat{A}$ transforms under a unitary operator $\hat{U}$ as: $\hat{A} \rightarrow \hat{A}' = \hat{U}^{\dagger}\hat{A}\hat{U}$. Thus, if $\hat{H}$ is invariant under the transformation corresponding to $\hat{U}$, then $\hat{H}$ must satisfy[10]

$$\hat{U}^{\dagger}\hat{H}\hat{U} = \hat{H} \longleftrightarrow [\hat{H}, \hat{U}] = 0. \tag{D.16}$$

If Eq. (D.16) is satisfied, then the observable corresponding to the generator of the transformation $\hat{U}$ is a constant of the motion (see Section 11.3.2).

Consider, for example, the unitary translation operator $\hat{T}$ (see Eq. (D.14)). If $[\hat{H}, \hat{T}] = 0$, then $\hat{T}^{\dagger}\hat{H}\hat{T} = \hat{H}$. Thus, $\hat{H}$ is translationally invariant along the x axis, and p_x—the observable corresponding to the generator $\hat{p}_x$—is a constant of the motion. That is, $d\langle p_x \rangle/dt = 0$.

This is an example of *Noether's theorem*, a famous result that's applicable in both classical and quantum mechanics. Noether's theorem states that, corresponding to each continuous symmetry transformation of a system there is a conserved quantity.[11] Referring to Eq. (D.14), we see that

1. $[\hat{H}, \hat{R}_z] = 0 \longrightarrow$ conservation of angular momentum in the z direction,
2. $[\hat{H}, \hat{V}] = 0 \longrightarrow$ conservation of x position,
3. $[\hat{H}, \hat{H}] = 0 \longrightarrow$ conservation of energy, unless $\hat{H} = \hat{H}(t)$.

---

[10] If $[\hat{H}, \hat{U}] = 0$, then $\hat{U}^{\dagger}\hat{H}\hat{U} = \hat{U}^{\dagger}\hat{U}\hat{H} = \hat{U}^{-1}\hat{U}\hat{H} = \hat{H}$.

[11] Noether's theorem in *classical* mechanics is discussed in: Hand (1998), pp. 170–175.

If, in addition to $[\hat{H}, \hat{R}_z] = 0$, we also have $[\hat{H}, \hat{R}_x] = 0$ and $[\hat{H}, \hat{R}_y] = 0$ (where $\hat{R}_x$ and $\hat{R}_y$ are rotation operators about the x and y axes, respectively), then total angular momentum is conserved.

For case 2 above, note that $\hat{x}$ is the generator of the transformation $\hat{V}$, while $\hat{p}_x^2/2m$ is the kinetic energy operator, and thus part of $\hat{H}$. But $[\hat{x}, \hat{p}_x]$ never vanishes, so this case is of little interest.

For case 3, $[\hat{H}, \hat{H}] = 0$ trivially. This means that the system energy is a constant of the motion, *unless* $\hat{H}$ depends explicitly on t.

In this book, our use of unitary operators is limited, apart from the time evolution operator. In some areas, however, such as particle physics, continuous symmetries and the related conservation principles form a central topic.

Bibliography

G. Arfken, *Mathematical Methods for Physicists* (Academic, New York, 1970)

R. Baierlein, Two Myths About Special Relativity, *American Journal of Physics*, **74**, 193–195 (2006)

L. Ballentine, Einstein's Interpretation of Quantum Mechanics, *American Journal of Physics*, **40**, 1763–1771 (1972)

L. Ballentine, *Quantum Mechanics: A Modern Development* (World Scientific, Singapore, 1998)

J. Bell, *Speakable and Unspeakable in Quantum Mechanics* (Cambridge University Press, Cambridge, 1987)

D. Bohm, *Quantum Theory* (Prentice Hall, New York, 1951), reprinted by Dover, New York, 1979

H. Boorse and L. Motz, eds., *The World of the Atom* (Basic, New York, 1966)

M. Born, *The Born-Einstein Letters* (Walker, New York, 1971)

P. Busch, "The time energy uncertainty relation," in *Time in Quantum Mechanics*, J. Muga *et al*, eds. pp. 69–98 (Springer-Verlag, Berlin, 2002)

A. Calaprice, *The Expanded Quotable Einstein* (Princeton University Press, Princeton, NJ, 2000)

J. Campbell, *The Inner Reaches of Outer Space* (Alfred van der Marck Editions, New York, 1985)

J. Cushing, *Quantum Mechanics: Historical Contingency and the Copenhagen Hegemony* (University of Chicago Press, Chicago, 1994)

P. Dirac, *Directions in Physics* (Wiley-Interscience, New York, 1978)

Einstein: A Portrait (Pomegranate Artbooks, Petaluma, CA, 1984)

R. Feynman *et al.*, *The Feynman Lectures on Physics, v. 3* (Addison-Wesley, Reading, MA, 1965)

A. Goldberg, H. Schey, and J. Schwartz, Computer-Generated Motion Pictures of One-Dimensional Quantum-Mechanical Transmission and Reflection Phenomena, *American Journal of Physics*, **35**, 177–186 (1967)

D. Griffiths, *Introduction to Quantum Mechanics, 2nd edn.* (Pearson, Upper Saddle River, NJ, 2005)

L. Hand and J. Finch, *Analytical Mechanics* (Cambridge University Press, Cambridge, 1998)

W. Heisenberg, "The physical content of quantum kinematics and mechanics," *Zeitschrift für Physik* **43**, 172–198 (1927); reprinted in *Quantum*

Theory and Measurement, J. Wheeler and W. Zurek, eds. (Princeton University Press, Princeton, 1983)

J. Hilgevoord and J. Uffink, "The uncertainty principle", in *The Stanford Encyclopedia of Philosophy* (Winter 2001 Edition), Edward N. Zalta, ed.; http://plato.stanford.edu/archives/win2001/entries/qt-uncertainty/

D. Howard, Who Invented the "Copenhagen Interpretation"? A Study in Mythology, *Philosophy of Science*, **71**, 669–682 (2004)

J.D. Jackson, *Classical Electrodynamics, 3rd edn.* (Wiley, New York, 1999)

M. Jammer, *The Conceptual Development of Quantum Mechanics* (McGraw-Hill, New York, 1966)

M. Jammer, *The Philosophy of Quantum Mechanics* (Wiley, New York, 1974)

T. Jordan, Why $-i\nabla$ is the Momentum, *American Journal of Physics*, **43**, 1089–1093 (1975)

P. Kosso, The Empirical Status of Symmetries in Physics, *British Journal of the Philosophy of Science*, **51**, 81–98 (2000)

L. Kovach *Boundary Value Problems* (Addison-Wesley, Reading, MA, 1984)

L. Landau and Lifshitz, *Quantum Mechanics* (Addison-Wesley, Reading, MA, 1958)

R. Liboff, *Introductory Quantum Mechanics, 4th edn.* (Addison-Wesley, San Francisco, 2003)

R. Loudon, *The Quantum Theory of Light, 3rd edn.* (Oxford Univerity Press, Oxford, 2000)

T. Maudlin, *Quantum Non-Locality and Relativity* (Blackwell, Oxford, 1994)

J. Mulvey, ed., *The Nature of Matter* (Oxford University Press, Oxford, 1981)

D. Park, *Introduction to the Quantum Theory* (McGraw-Hill, New York, 1964)

B. Pullman, *The Atom in the History of Human Thought* (Oxford University Press, Oxford, 1998)

J. Rigden, Editorial: Heisenberg, February 1927, and physics, *American Journal of Physics*, **55**, p. 107 (1987)

H. Robertson, The Uncertainty Principle, *Physical Review*, **34**, 163–164 (1929)

J. Sakurai, *Modern Quantum Mechanics, revised edn.* (Addison-Wesley, Reading, MA, 1994)

J. Sayen, *Einstein in America* (Crown, New York, 1985)

L. Schiff, *Quantum Mechanics* (McGraw-Hill, New York, 1968)

P. Schilpp, *Albert Einstein: Philosopher-Scientist* (Open Court, La Salle, IL, 1949)

R. Shankar, *Principles of Quantum Mechanics* (Plenum, New York, 1980)

J. Taylor, *An Introduction to Error Analysis, 2nd edn.* (University Science Books, Sausalito, CA, 1997)

J. Townsend, *A Modern Approach to Quantum Mechanics* (McGraw-Hill, New York, 1992), reprinted by University Science Books, Sausalito, CA, 2000

A. Watts, *The Way of Zen* (Vintage, New York, 1989)

J. Wheeler and W. Zurek, eds., *Quantum Theory and Measurement*, (Princeton University Press, Princeton, NJ, 1983)

Index

Printed in the United States
By Bookmasters